SpringerBriefs in Pharmacology and Toxicology

SpringerBriefs in Pharmacology and Toxicology present concise summaries of cutting-edge research and practical applications across a wide spectrum of fields.

More information about this series at http://www.springer.com/series/10423

Soodabeh Saeidnia • Ahmad Reza Gohari
Azadeh Manayi • Mahdieh Kourepaz-Mahmoodabadi

Satureja: Ethnomedicine, Phytochemical Diversity and Pharmacological Activities

Soodabeh Saeidnia
Medicinal Plants Research Center
Tehran University of Medical Sciences

Azadeh Manayi
Medicinal Plants Research Center
Tehran University of Medical Sciences

Ahmad Reza Gohari
Medicinal Plants Research Center
Tehran University of Medical Sciences

Mahdieh Kourepaz-Mahmoodabadi
Medicinal Plants Research Center
Tehran University of Medical Sciences

ISSN 2193-4762
ISSN 2193-4770 (electronic)
ISBN 978-3-319-25024-3
ISBN 978-3-319-25026-7 (eBook)
DOI 10.1007/978-3-319-25026-7

Library of Congress Control Number: 2015953447

Springer Cham Heidelberg New York Dordrecht London

Printed on acid-free paper

Springer International Publishing AG Switzerland is part of Springer Science+Business Media (www.springer.com)

Preface

Among all the medicinal plants growing around the world, some species have attracted a great concern due to their capability of producing a broad spectrum of bioactive natural products as well as their evidence-based pharmacological activities. *Satureja* (trivial name: Savory) belongs to the Lamiaceae family as one of those mentioned plants. This genus comprises about 200 species worldwide, which are mostly aromatic herbs and shrubs with numerous therapeutic effects representing considerable diversity in their chemical composition and biological properties as well as medicinal effects. Although we can find a few review articles or books (mainly in non-English language) to include some species of this genus, there is no comprehensive book or review especially to gather all the useful information on "ethnomedicine and traditional usage; microscopic characterizations; chemical diversity; pharmacology and biological activities." The authors of the present book have been involved in different studies on various species of this genus, growing in Iran, for many years, and published several research articles thereof. Therefore, we believe this is a suitable time to publish a comprehensive book on Iranian species of *Satureja* to conclude most of the above mentioned aspects.

Researchers in the field of pharmaceutical sciences and natural medicines, pharmaceutical companies who produce herbal/natural products, students in the field of pharmacognosy and phytochemistry, academic scientists in the mentioned fields, as well as those who work on the areas of traditional medicine and pharmacy might be interested in getting botanical, morphological, pharmacological and phytochemical information about these valuable medicinal species, which is concluded briefly in the present book.

I would like to thank the contributors of the chapters, who are my great colleagues, for their kind endeavors in creating this text. Moreover, I would like to acknowledge the support of the Springer staff, especially former staff member Manika Power. Any comments and feedback from the experts in the field of this book are welcome and will be considered for a future edition.

Soodabeh Saeidnia (Pharm.D., Ph.D.)

Contents

Chapter 1
Introduction

The genus, *Satureja* (Savory), belongs to the well-known plant family Labiatae (Lamiaceae), subfamily Nepetoidae, tribe Mentheae and comprises about 200 species worldwide. These are mostly aromatic herbs and shrubs distributed widely in Middle East, the Mediterranean area, West Asia, North Africa, Canary Islands, and boreal America [1–5]. About 30 species of this genus are commonly named savory, among which summer savory (*Satureja hortensis*) and winter savory (*Satureja montana*) are mainly cultivated [6]. Some species of this genus like *S. hortensis* have been used as the herbal tea and food additive [7]. In addition, the plant has traditionally been used in Iran for treatment of cramps, muscle pains, nausea, indigestion, diarrhea, and infectious because of its anti-spasmodic, anti-diarrheal and antimicrobial properties [8, 9]. Also, these plants were employed for treatment of some human disorders. For instance, *S. boliviana* has been applied in colds, diarrhea and stomach pain. Moreover, *S. parvifolia* has been reported to be applied against fever, rheumatic pains, stomachache, dyspepsia, gastrointestinal bloating, diarrhea, influenza, and colds [10, 11]. Other preparations such as infusions of *S. thymbra* leaves were also used for reduction of blood pressure, pain in joints, and antimicrobial activity [12]. In a number of old books about Persian Traditional Medicines, some beneficial effects are mentioned for internally application of savory including appetizer, anti-cough, strengthen of eye, anti-vomiting agent, reducing toothache and externally for relieving rheumatic pain and inflammation. Moreover, decoctions of the plant have been used in treatment of scabies and itching. Also, a preparation of the savory flowers is commonly used as emmenagogue and diuretic. It is also indicated that the savory seed is helpful for treatment of tooth ache, joint ache, and hemorrhoid [13]. There is just one review article updating pharmacology of *Satureja* species until May 2010 principally focused on their pharmacological activities [1]. Therefore, in this book, we aimed to present an over review of the main secondary metabolites in various species of *Satureja* as well as the important biological and pharmacological activities. Underlying mechanism of action for most of the medicinal properties, ethnobotany and micromorphological characterizations and also clinical studies on this valuable plant genus have been other subjects of this book.

S. Saeidnia et al., *Satureja: Ethnomedicine, Phytochemical Diversity and Pharmacological Activities,* SpringerBriefs in Pharmacology and Toxicology,
DOI 10.1007/978-3-319-25026-7_1

Botanical characterizations of the genus, *Satureja*, natural habitats, and different species of *Satureja* have been concisely described in the present chapter. The genus *Satureja* L. (savory) belongs to the Lamiaceae family with 200 species of herbs and shrubs, in which 16 species are growing in Iran. These plants mostly grow in the Mediterranean region, Europe, West Asia, North Africa, the Canary Islands, and South America [1]. This genus includes annual and perennial shrubbery plants with several woody branching stems. As literature reveals, 16 species (including *S. atropana, S. laxiflora, S. macrosiphonia, S. spiciger, S. sahendica, S. bachtiarica, S. isophylla, S. kallarica, S. khuzistanica, S. boissieri, S. mutica, S. macrantha, S. intermedia, S. rechingeri, S. edmondi*, and *S. kermanshahensis*) are endemic of Iran. Geographical distribution of different *Satureja* species in Iran together with their world distribution has been described in Table 1.1. They are mainly distributed in the west parts of Iran. The exceptions are *S. isophylla, S. boisseri, S. mutica*, and *S. spicigera*, which grow in the north parts of Iran [2, 3].

The botanical characterizations of the most important species of *Satureja*, particularly those are exclusively growing in Iran, have been described below. The followed botanical descriptions were extracted from two Iranian botanical sources: Jamzad (2012) and Maroofi (2010) [2, 3] including some Latin botanical expressions.

1.1 *Satureja spicigera* (C. Koch) Boiss.

The mentioned species is a perennial woody herb at the base with 25–60 cm height. Stem: slender, warp and widespread; Leaf: densely, numerous branches at the base, herbaceous, green light on both surface, densely hairs and sessile, secretory in the lower surface, attenuate at the base into petiole, acute, median nerve slender, lower leaves 15–20 mm length and 2–4 mm in weight, upper leaves and leaves of branches small, 1 mm, verticillasters numerous; Flower: three flowered, spike, slender peduncle, calyx 3–4.5 mm, campanulate-funnel, sparse canescent or sessile, sub-bilabiate, lower slightly longer, lanceolate, upper teeth 1/3 length of calyx; Corolla: 8–10 mm, almost white- pink, corolla closet, stamens and styles exerted; Fruit: nutlets, almost spherical, wide. Flowering period is in autumn.

1.2 *Satureja mutica* Fisch. and C. A. Mey.

This plant is also perennial and woody at the base with 30–50 cm height. Stem: numerous, branched from the base, branches slender, with grey and short hairs; Leaf: lower leaves 30 × 5 mm, attenuate at the base into petiole, acute, flat, secretory, upper leaves and leaves of branches small, narrow, oboval-lanceolate, rounded; Flower: verticillasters numerous, three flowered, short pedicel, linear bracts, lax, shorter than the calyx, calyx 5 mm, hisped-glandulary hairs, bilabiate, calyx teeth almost linear, uppers 1 mm and lowers 2 mm; Corolla: 7 mm; Fruit: nutlets 1–1.5 mm length and 1 mm width, obovate rounded. Flowering period is in autumn.

Table 1.1 Some information about *Satureja* species growing in Iran

Name	Flowering period	Distribution in Iran	Distribution in Iran province	World distribution	Synonym
S. intermedia C. A. Mey., Verz. Cauc.	Summer	North West	Azerbaijan, Ardebil, Kerman	Ghafghaz, Iran	–
S. atropatana Bunge[a]	Autumn	North West	Azerbai-jan, Tabriz (endemic)	Iran	–
S. macrantha C. A. Mey	Autumn	West and North West	Tabriz, Zan-jan, Hamedan, Kermanshah	Ghafghaz, Iran, Iraq	–
S. laxiflora C. Koch	Last summer and early autumn	West and North West	Azerbaijan, Tehran, Ghazvine	Ghafghaz, Iran, Turkey	*S. hortensis* L. var. *Laxiflora, S. hortensis, S. grandiflora Boiss.*
S. sahendica Bornm[a]	Autumn	West and North West	Azerbaijan, Kurdistan, Kermanshah	Iran	–
S. bachtiarica Bunge[a]	Autumn	West and Center	Kermanshah, Isfahan, Yazd, Kohgiluyeh and Boyer Ahmad, Bakhtiari, Fars, Kerman	Iran	–
S. isophylla Rech. f.[a]	Autumn	North	Mazandaran	Iran	–
S. kallarica Jamzad[a]	Summer	West	Bakhtiari	Iran	–
S. khuzistanica Jamzad[a]	last summer and early autumn	South West	Lorestan, Khuzesatn	Iran	–
S. rechingeri Jamzad[a]	Autumn	South West	Ilam	Iran	–
S. edmondi Briquet[a]	Autumn	West	Kermanshah, Lorestan, Bakhtiari	Iran	*S. longiflora* Boiss. and Hausskn
S. macrosipho-nia Bornm	Autumn	West	Lorestan	Iran, Iraq	–
S. kermansha-hensis Jamzad[a]	Autumn	West	Kermanshah	Iran	–
S. boisseri Hausskn. ex Boiss.	Autumn	North	Gorgan, Gillan	Iran, Turkey	–
S. mutica Fisch and C. A. Mey	Autumn	North and North East	Mazandaran, Kharasan	Ghafghaz, Iran, Central Asia	–

Table 1.1 (continued)

Name	Flowering period	Distribution in Iran	Distribution in Iran province	World distribution	Synonym
S. spicigera (C. Koch) Boiss.	Autumn	North	Gorgan, Gil-lan, Tehran	Ghafghaz, Iran, Turkey	*Micromeria spicigera* C. Koch, *S. intermedia* C. A. Mey *Lax-ior* Benth., *S. alternipilosa* C. Koch, *Micromeria alternipilosa* C. Koch

[a] The endemic species that is exclusively growing in Iran

1.3 *Satureja boissieri* Hausskn

The species is perennial herb with 40–60 cm height. Stem: narrow, simple and without divergence, descending curved hairs; Leaf: 10–26 mm length, 2–5 mm weight, linear, obtuse, upper linear, canaliculate, canescent short-curved hairs and glandular in teo surface. Verticillasters: 3–8 mm, cluster inflorescence 8–20 cm length and 15–30 mm diagonal, calyx 4–5 mm, subbilabiate, upper teeth 0.75–1 mm and lower 1.5–2 mm, villous; Corolla: 9 mm, corolla tube exerted, stamens exerted; Fruit: nutlets 1.2–1.5 mm length and 1 mm weight, dark brown. Flowering period is in autumn.

1.4 *Satureja macrosiphonia* Bornm

This plant is perennial too, with 30–40 cm height. Stem: branches at the base and at the upper parts, very thin branches, crinkle at the end with small leaves, along with smaller leaves, canescent short-white hairs, curved or erect in retrorse; Leaf: 3–5 mm, linear, lower leaves more or less smooth; Bractes: 1–2 mm length and 0.5–0.6 mm width, elliptic-oblong, 1–4 flowered, peduncle 3–6 mm, calyx 3 mm, campanulate-tubular, upper teeth triangular 1 mm, lower teeth lanceolate-linear 1.5 mm, subbillabiate; Flower: 14 mm, pink with violet edge, long tube, tube exerted from calyx, upper lip of corolla have a short cut in the middlelower lips equal 3-lobled, front stamens almost flat corolla throat; Fruit: nutlets 1.5–1.7 mm length and 1 mm weight, obovate. Flowering period is in autumn.

1.5 *Satureja atropatana* Bunge

Satureja atropatana is a perennial herb, woody at the base, finely papillous, 30–75 cm long. Stem: numerous, erect, often simple, divaricate, glabrous on the nodes, leafy; Leaf: small, oboval-oblong, upper linear, canaliculated-concave, obtuse, florals short; Flower: violet-pink, verticillasters more or less multi-flowered, remote, shortly peduncled or sessile, bracteoles numerous; Calyx: 5 mm long, subbilabiate, tubular, laxly papillous-hirsute, upper teeth shortly deltoid-triangular, lowers slightly longer, lanceolate, obtuse, four times shorter than the tube; Corolla: exerted, 12 mm long; Fruit: nutlets ovoid-oblong, pale, smooth. Flowering period is usually June–July.

1.6 *Satureja laxiflora* C. Koch

This species is dwarf annual plant with 10–20 (–30) cm height. Stem: branched from the base, branches numerous, lax, filiform, erect or spreading; Leaf: 3×15 mm, narrowly linear, acute, attenuate into petiole; Flower: pink; verticillesters 1–2 (–3) flowered; peduncles 5–15 (–50) mm long, axillary; lower flower in each glomerlules peduncled, other sessile; Calyx: 3–3.5 mm long, regular, campanulate, rigid, villous, teeth equaling the tube or longer, lanceolate- subulate; Corolla: 8–10 mm long, pubescent, tube exerted, upper lip shortly 2-lobed; Stamens: two stamens, inserted under the lower lip; Fruit: nutlets 1×0.5 mm, rounded-ovoid, 4-nerved. Flowering period is during June–July.

1.7 *Satureja macrantha* C. A. Mey

It is well-characterized by a perennial herb, woody at the base and 30–50 cm height. Stem: numerous, canescent, scabrous, simple or branched; Leaf: 8–15 (–25) $\times$ 2–3 mm, thickened, linear-spathulate, attenuate at the base, sessile, obtuse or acuminate, spreading, facicled-axillary, green-canescent, flat, median nerve and margin glandular-punctate, finely scabrous-puberulent; floras short, narrowly, linear, attenuate towards apex; Flower: pink; verticillasters few-flowered (1–3 flowers); Calyx 5–6 mm long, tubular, hairy, campanulate, bilabiate, teeth triangular, subulate, muticous, upper three times shorter than the tube, lowers long and narrow; Corolla: three times longer than the calyx, 10–15 mm long, tube elongate; upper lip slightly emarginated, lowers unequal, 3-lobed; Stamens and styles are shortly exerted; Fruit: nutlets 1.5×0.75 mm, ovoid. Flowering period is during May–June.

1.8 *Satureja intermedia* C. A. Mey.

This species is also perennial with 10–20 cm height, woody and thicken at the base. Stem: numerous, simple, curved-ascending; Leaf: dense, white canescent hairs; Lower leaves with petiole 8–10 mm length and 4–6 mm width, thick median nerve, flattened, oboval-oblong, secretory. Flower: 3–5 flowered, thyrsu, radial, lax; Bractes: linear, limber, shorter than the calyx; Calyx: 5–8 mm, subbilabiate, upper teeth lanceolate-subulate 1.5–2.5 mm, lower teeth lanceolate-linear 2–3.5 mm, corolla red 8–10 mm, corolla tube hide in calyx, lower lip some longer than the upper lip; Stamens: exerted more or less from flower tube: Fruit: nutlets brown, 1.5–1.75 mm length and about 1 mm width, flat surface and divaricate. Flowering period is in summer.

1.9 *Satureja sahendica* Bornm.

The plant is perennial. Stem: numerous, 12–25 cm high, tenuos, ascending, divaricate at the base; Leaf: dense, curved hairs in the bottom part, 5–12 mm length and 1–3 mm width, attenuate at the base, sessile with acuminate more or less obtuse, fold, thick tissue, grey, white short-canescent hairs; Flower: 2–6 flowers, peduncle 1 mm long; Calyx: 3–6 mm, tubular-campanulate, bilabiate, grey, upper teeth short and triangular at the base, about 0.5 mm, lower teeth slightly longer, thin; Flower: 6–12 mm, white or violet-pink, villous, two times longer than the calyx, short edge; Stamens; two perior stamens are longer, exerted from corolla tube; Fruit: nutlets 1.5 mm length and 0.5 mm width, oblong, dark brown. Flowering period is in autumn.

1.10 *Satureja bachtiarica* Bunge.

It is a perennial species, 20–45 cm height. Stem: numerous, short branches, with grey short hairs, erect, thick; Leaf: lower leaves 5–10 mm length and 1.5–3.5 (–4) mm width, oblong-spathulate or oblong-linear, thick, sessile, re-curved conduplicate, glands on both sides, grey, white stiff hairs; Flower: verticillasters, many flowered, calyx 1.5–3 mm, campanulate, teeth almost unequal, lanceolate-triangular, hairy with sessile glands, longer teeth almost equal with tube length, flower white; Corolla: 3–6 mm, villous, corolla tube hide more or less inside calyx, stamens exerted; Fruit: nutlets 1.4–1.7 mm length and 0.5–0.7 mm width, obovate-oblong, flat. Flowering period is in autumn.

1.11 *Satureja isophylla* Rech. f.

The plant is perennial. Stem: numerous, slightly, dense, 6–10 cm, short branches, short canscent hairs; Leaf: numerous, same scale, 6–(8)–10 mm length and 1.5–3 mm width, attenuate at the base, lanceolate slender, green grayish; Flower: verticillaster, few flowers, calyx 3–4 (–5) mm, subbilabiate, light green, canescent, teeth broad at the base and immediate attenuate, subulate,1/3 or 1/4 calyx tube; Corolla: 10 mm, white, exerted from calyx. Upper lip corolla short, erect, lower lip slightly longer from the upper lip, 3-lobed more or less equal; Stamens: exerted the flower. Fruit: 1.5–2 mm length and 0.5 mm width, oblong, light brown. Flowering period is in autumn.

1.12 *Satureja kallarica* Jamzad

The plant is perennial, curved or ascending, 15–20 cm, branches, villous; Leaf: 4–11 mm length and 3–6.5 mm width, shortly petiolate 1.5 mm, obovate limb; Flower: 2–5 flowered, linear bractes, peduncle 2–4 mm; Calyx: 7–8 mm, tubular, spreading hairs, subbilabiate with teeth almost equal, straight 1.5 mm; Corolla: 12 mm, white, straight tube, upper lip straight, lower lip with 3-lobed; Stamens: four stamens, didinamous exerted from tube.; Fruit: nutlets 1.2 mm, oblong. Flowering period is in summer.

1.13 Satureja *khuzistanica* Jamzad.

This species is perennial with about 30 cm high. Stem: divaricate, short hairs and glands; Leaf: dense, short internode 2–3 mm, alternative, flattened until tube, 6–8 mm length and 3–5 mm width, gyrate- obovate, slender at the base, covered with hairs, dense in underside, dense glands on upside; Flower: verticillasters 2–8 flowered, shortly peduncle, lanceolate bractes, peduncle 0.5–1 mm, calyx 5.5–6 mm, tubular-campanulate, bilabiate, upper teeth triangular 1 mm, lower teeth linear 1.5 mm; Corolla: 11 mm, violet, bilabiate; Stamens: four, lower twin more or less exerted of flower tube; Fruit: nutlets 2–2.1 mm length and 1–1.1 mm width, obovate. Flowering period is autumn-winter.

1.14 *Satureja rechingeri* Jamzad

The above mentioned species is perennial with woody base, divaricate at the base. Stem: about 50 cm, covered with long-gray hairs; Leaf: 9–13 mm length and 7–11 mm width, broad obovate or gyrate, gently slender at the base, dense gland, lower leaves flattened, upper gently at top conduplicate; Flower: peduncle 2–6 mm, 2–8 flowered, spike, lanceolate; Bractes: 2–3.5 mm; Calyx: 6.5–8 mm, tubular-campanulate, bilabiate, upper teeth triangular 2 mm, acuminate, lower teeth lanceolate 3 mm; Corolla: 12–15 mm light, yellow, violet in lip, and stamens exerted of flower. Flowering period is in autumn.

1.15 *Satureja avromanica* Maroofi

This is a suffruticose perennial herb with 35–80 cm height, several stemmed, slender, mostly simple or with a few branches, ascending-arcuate to erect, grayish green above; Leaves: opposite or in fascicles, entire, lax, sessile or subsessile in lower part with petiole up to 1 mm long, cuneate-oblong to cuneate-obovate or lanceolate, up to 35 mm long, upper leaves smaller than the lowers, grayish-green; Flowers: pedicels 1–4 mm long, shorter than the calyx, bractes oblong-ovate, two bracteoles, smaller than the bractes, nearly the same shape. Calyx pubscent, 3.5–6.5 mm long, tubular, 2-labiate, teeth of the calyx shortly hairy at the margin; lower lip teeth 1.7–2 mm, subulate-triangular; Corolla: pubscent, slender, straight; Stamens: four, included in tube; upper filaments 1.5 mm and lower filaments 1.8–2.5 mm long; Four nutlets, minutely glandular-hairy above, with obtuse-rounded apex [3].

Chapter 2
Satureja: Ethnopharmacology and Ethnomedicine

2.1 Introduction

In the Persian traditional books (written in Persian or Arabian languages), some beneficial effects are mentioned for internal application of savory including appetizer, antitussive, strengthen of eye, anti-vomiting agent, reducing tooth ache and externally for relieving rheumatism pain and inflammation. Moreover, decoction of the plant is used in treatment of scabies and itching. A preparation of the savory flowers is commonly used as emmenagogue and diuretic. It is also indicated that the savory seed is helpful to treat tooth ache, joint ache, and hemorrhoid.

2.2 Natures in Iranian Traditional Medicine

In traditional Iranian medicine all things including alive or not alive are created form four elements including:

1. Fire which is hot
2. Water that is wet
3. Earth or soil that is dry
4. Air that is cold

Nature or temperament of animals and human is a result of mixing four special elements including:

1. Blood that is hot and wet
2. Bile that is hot and dry
3. Lymph or phlegm that is cold and wet
4. Atrabilious that is cold and dry

In the Fig. 2.1, a schematic diagram of human nature is shown as indicated in Persian Traditional Medicine. Any imbalance in the elements of a nature can be resulted in creation of a disease [14].

S. Saeidnia et al., *Satureja: Ethnomedicine, Phytochemical Diversity and Pharmacological Activities*, SpringerBriefs in Pharmacology and Toxicology,
DOI 10.1007/978-3-319-25026-7_2

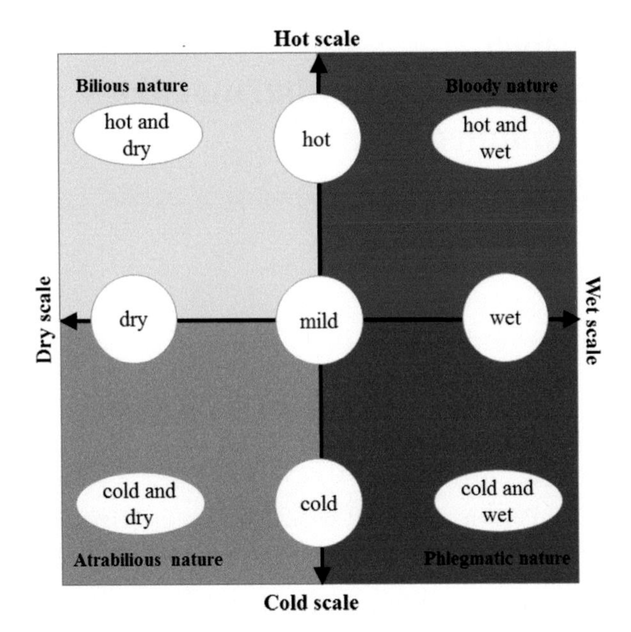

Fig. 2.1 Schematic diagram of human natures in Persian traditional medicine

2.3 Traditional Applications of Savory

The brief information of the above mentioned temperaments in Iranian traditional medicine might be helpful to understand the traditional usage of savory plants. According to the Iranian traditional literature, summer savory (*Satureja hortensis*) is hot and dry in its nature [15]. It is indicated that savory possesses appetizing activity and strengthen the potency of males. It is reported that a mixture of dried summer savory with equal amount of sugar strengthens eye power and prevents eye drip. Traditionally, decoctions of the plant have a number of beneficial effects in fluctuation, diarrhea, vomiting and hearth pain, as well as tooth ache. Also, we can find that a decoction of summer savory with celery juice is helpful in breakage of kidney stones and dysuria. Moreover, a mixture of figs and savory cures asthma or coughs and brightening skin color of face. In the situation of uvula inflammation, coughs, and scorpion bite, a mixture of summer savory with honey can be effective. Furthermore, a plaster of the plant on the skin can be used to treat scabies. In addition, a mixture of summer savory juice with milk is mentioned to be employed in ear ache. Rheumatoid and nervous pains as well as itching and strengthen of weak children are some examples of other disorders that might be cured with tacking bath using decoction of summer savory. It is reported that an external paste of summer savory powder with honey reduced the back pain, sciatic, or discopathy [15].

There are some beneficial effects related to the flowers of the plant. A powder of flowers with vinegar and salt is diuretic and emmenagogue orally. Chewing of summer savory seeds is appetizer and reduces tooth ache. The paste of the seeds also reduces pain in joints inflammation [15].

Chapter 3
Micromorphological Characterizations

Botanical microscopy is a valuable procedure for identification and quality assurance of herbal ingredients as well as an inherent part of almost all pharmacopoeias. In this section, the microscopic characterizations of some *Satureja* species have been described. Actually, different powders of the plants were decolorized for better observation before microscopic observations.

3.1 Experimental Procedure

One gram of each tissue powder (leave, flower, stem and root) of *Satureja* species was separately boiled in potassium hydroxide solution (10%) in a backer on heater for 30 s (or 1 min), depending on the tissue hardness, and washed afterwards with distilled water three times. The powders were successively treated with sodium hypochlorite for bleaching and then washed with distilled water. The preparation was mounted in aqueous glycerin. Photomicrographs were taken using Zeiss microscope attached with a digital camera. Photomicrographs of the sections were taken at different magnifications depending upon the microscopic details that needed to be observed. The slides were stained with specific stains such as methylene blue, toluidine blue, Sudan red G (for essential oils, resins, fats, and fatty oils), and then observed under the microscope [6].

Some structures are critical in identification of the plant powders. Such structures include stomata, trichomes, cicatrix, pollen grains, and crystal shapes. The mentioned structures are in different shapes in various plant powders that provide an excellent support in accurate identification. Stomata are pores in the epidermis of the different parts of the plants, through which exchange of gases and water takes place. Stomata have two subsidiary cells and according to the position of subsidiary cells with the guard cells (two kidney-shape cells that regulate closing and opening of stomata) different type of stomata are distinguished. Moreover, subsidiary cells are different in shape and size from other epidermal cells. A certain family of plants usually contains one type of stomata. Trichomes are cells or groups of cells

S. Saeidnia et al., *Satureja: Ethnomedicine, Phytochemical Diversity and Pharmacological Activities*, SpringerBriefs in Pharmacology and Toxicology,
DOI 10.1007/978-3-319-25026-7_3

that project as hairs from epidermal surface. Cicatrix is characteristic scar represent a place that trichomes break off from epidermal cells. Crystals of calcium oxalate in the plant cells are also provided a highly effective diagnostic feature in herbal powders. In a plant powder, some parts like epidermis, cork, fibers, and vessels typically remain intact but delicate tissues like trichomes, cambium, and sieve cells may disintegrated and hardly identified. Based on American Herbal Pharmacopiea, vessels usually are not suitable for differentiation between plants powders [6]. Microscopical characterization of different parts of some *Satureja* species, which are growing in Iran, including *S. bachtiarica*, *S. hortensis*, *S. atropana*, *S. macranta*, and *S. sahenidca* have been observed by microscope and illustrated followed.

3.2 *S. bachtiarica* Bunge.

3.2.1 Leaf

As it is indicated in Fig. 3.1, the epidermis of leaf powder consists of the cells with anticlinal walls and diacytic stomata. Stomata have two subsidiary cells and the mentioned cells show a common wall at the right angles to the longitudinal axis of the guard cells. This kind of stomata is typical in Lamiaceae family [6].

Glandular trichomes with multicellular glandular scales indicated a short stalk and usually ten secretory cells and detached coticule (Fig. 3.2a). This type of glandular trichome is often found in the members of Lamiaceae family, and therefore can be helpful in identification of unknown plant material. The glandular trichomes become orange-red in the presence of Sudan red that is an indicator for essential oil in the cells (Fig. 3.2b).

Uniseriate covering trichomes, known as non-glandular trichomes, together with warty coticule are abundant in the leaf segments (Fig. 3.3). This kind of trichomes stained orange when the leaf powder had already been treated with Saudan red indicating that the covering trichomes contain oil (Fig. 3.3c).

Branching non-glandular trichome is another type of trichome with acute tips that have been found in the leaf segments of *S. bachtiarica* (Fig. 3.4).

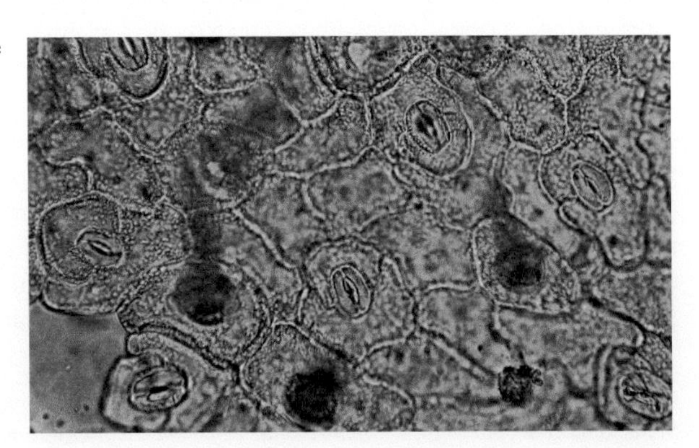

Fig. 3.1 The epidermis in the leaf powder of *S. bachtiarica*, showing diacytic stomata

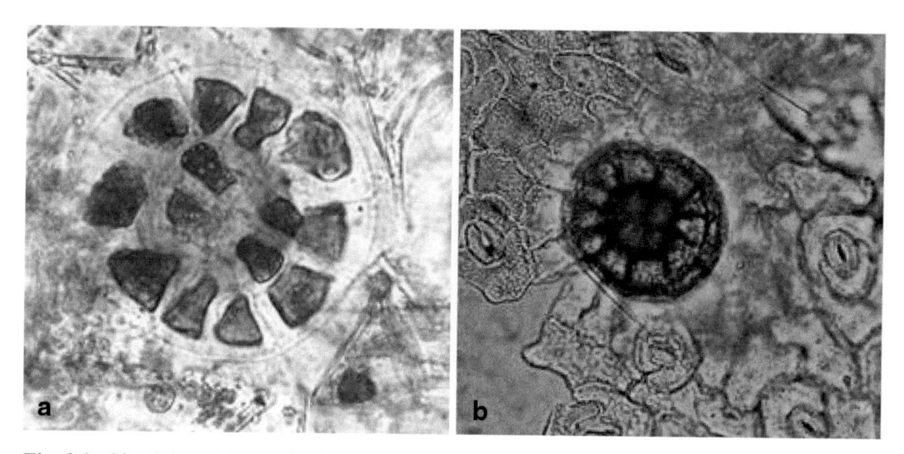

Fig. 3.2 Glandular trichomes in the leaf powder of *S. bachtiarica*; **a** treated with *methylene blue*; **b** treated by *Sudan red*

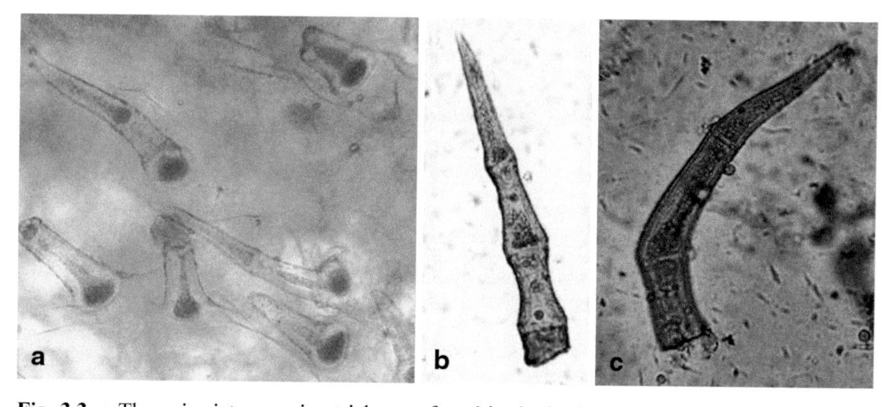

Fig. 3.3 **a** The uniseriate covering trichomes found in the leaf powder of *S. bachtiarica*; **b** one uniseriate covering trichome treated with *methylene blue*; **c** a uniseriate covering trichome treated with *Sudan red*

Fig. 3.4 A branching trichome in the leaf powder of *S. bachtiarica*

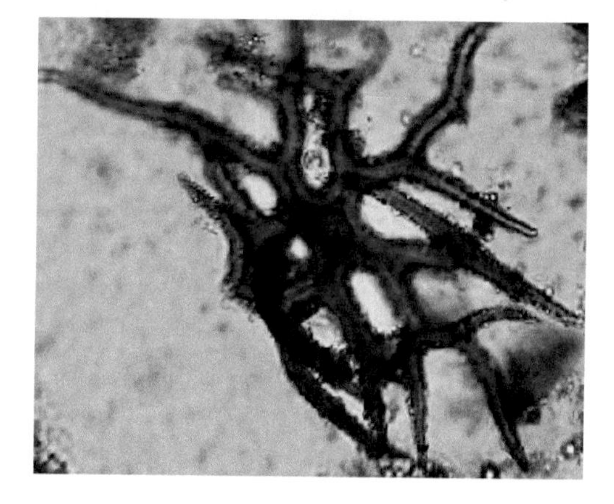

In the leaf segments of this species, another kind of trichome was identified with acute tips (Fig. 3.3) shorter than those belong to other type of covering trichomes (Fig. 3.3). This kind of trichome changed to orange-red in the presence of Sudan red (Fig. 3.5b) indicating oil accumulation in them. These covering trichomes contain cystoliths at the base, which are inorganic concentrations usually calcium carbonate in epidermal cells.

The epidermis of the leaf powder in *S. bachtiarica* showed some cicatrixes as the scars that are created by trichome cleavage from the surface of the plant. In the Fig. 3.6, this scar in the epidermis of the leaf powder of the plant is exhibited as a round white hole that splits apart with a radiating layer of enlarged cells.

A number of fragments of annularly and spirally thickened veins attached to the fibers (Fig. 3.7b) have been identified in the powder of the plant leaf.

3.2.2 Flower

Microscopic characteristics of the floral parts of *S. bachtiarica* represent a stamen consisting of a filament (Fig. 3.8a) and an anther (Fig. 3.8b), which are detached. This structure is useful for identification of the floral materials in comparison to other plant parts.

In the fragments of the flower powder of the plant, an ovary was also characterized that developed to fruit (Fig. 3.9).

Calyx of the plant densely covered with short unicellular and bicellular covering trichomes similar to those found on the leaf. Glandular trichome (specified with black arrow in the Fig. 3.10) has also been identified on the calyx.

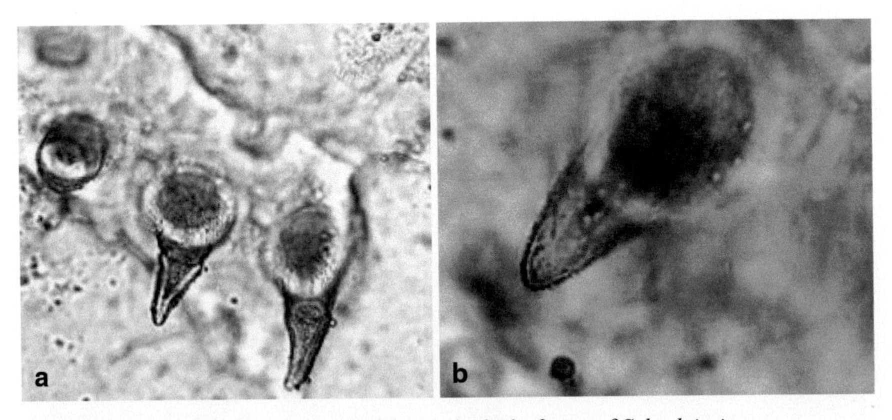

Fig. 3.5 The short uniseriate covering trichomes in the leaf parts of *S. bachtiarica*

Fig. 3.6 Epidermis of the leaf of S. bachtiarica containing stomata and cicatrix resulted from broken covering trichome.

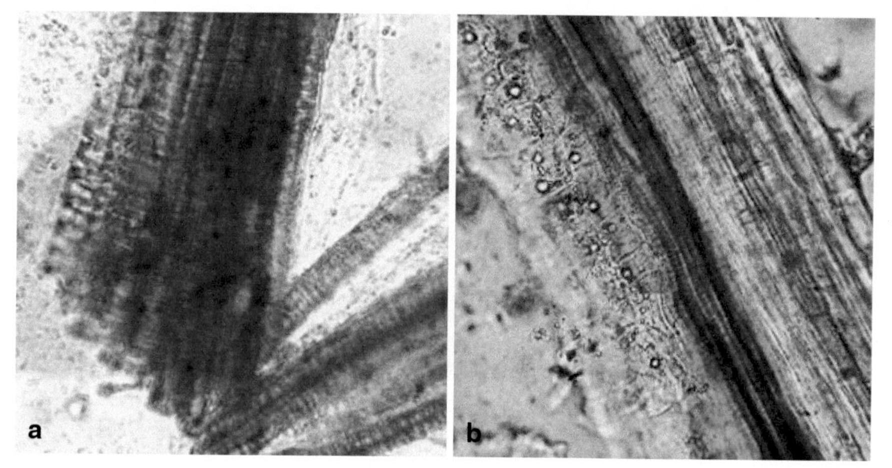

Fig. 3.7 **a** A number of veins in the leaf of *S. bachtiarica*; **b** fragments of the vein attached to the fibers

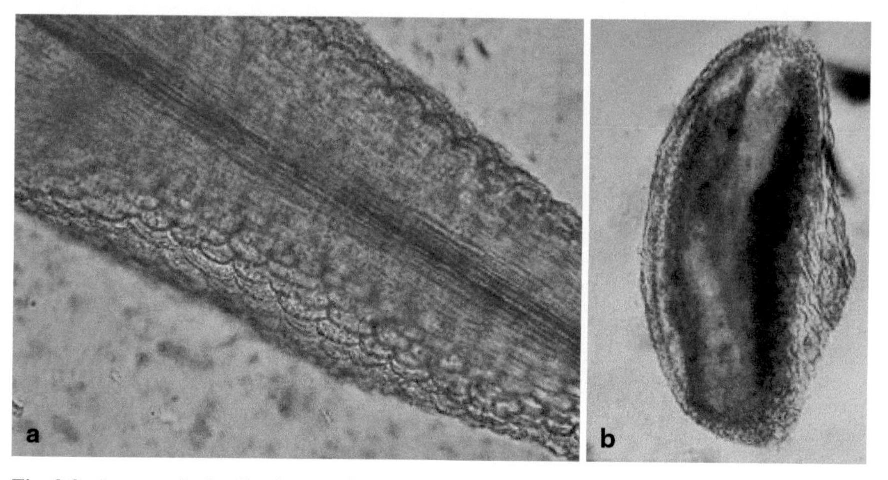

Fig. 3.8 A stamen in the floral parts of *S. bachtiarica*; **a** filament contains vessels; **b** surface view of anther

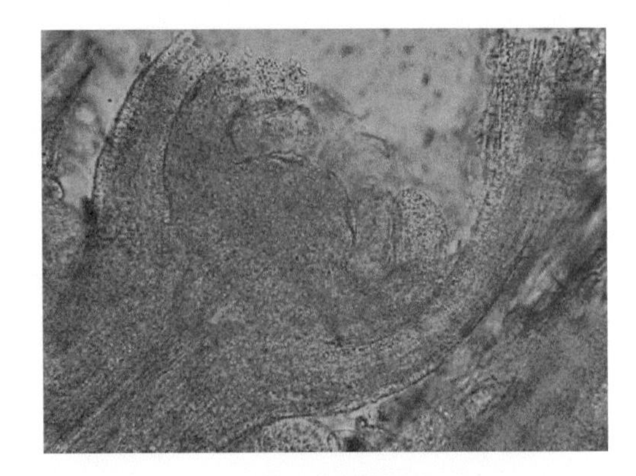

Fig. 3.9 A structure of ovary in the floral parts of *S. bachtiarica* in the surface view

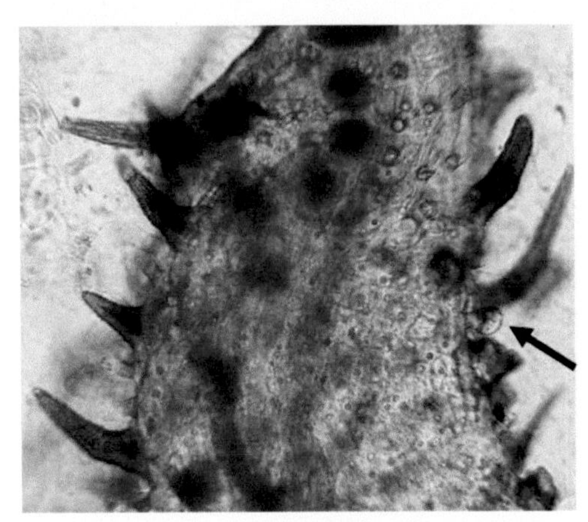

Fig. 3.10 A calyx of *S. bachtiarica* in the surface view shows covering trichomes and glandular trichome

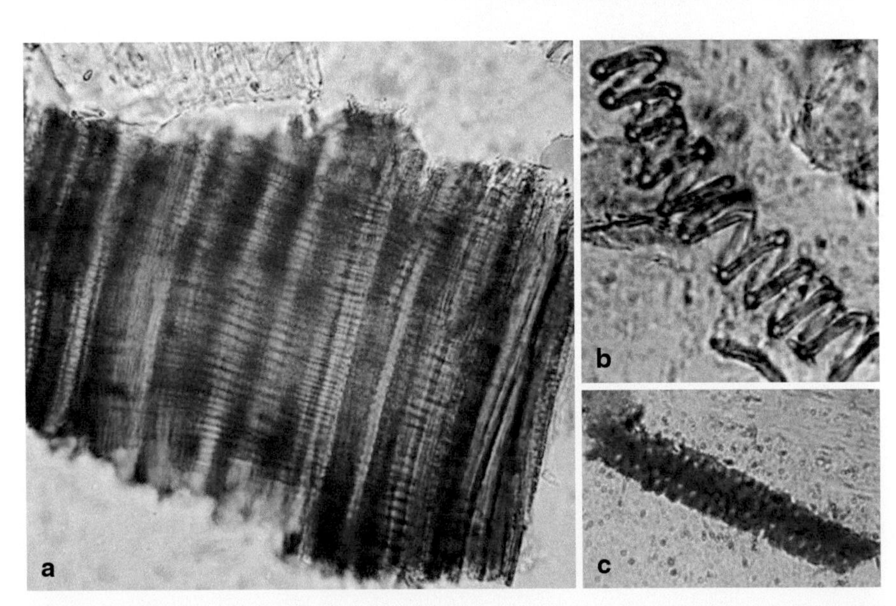

Fig. 3.11 Vessels in stem segments of *S. bachtiarica*; **a** group of vessels with fiber; **b** a vessel with spiral thickening; **c** a pitted vessel

3.2.3 Stem

A number of vessels in the stem segments of *S. bachtiarica* were found in single and/or large pieces with fibers (Fig. 3.11a). Some of them occur with spiral thickening (Fig. 3.11b). A fragment of the bordered pitted vessel was also observed (Fig. 3.11c).

A uniseriate covering trichome with a short basal cell and elongated terminal cell was found detached from epidermis in the stem powder of *S. bachtiarica*. It shows conical shape with slightly thickened cell wall and may be slightly swollen at base (Fig. 3.12). This covering trichome is similar to those found in other parts of the plant powder like leaf and calyx.

The epidermis in the stem powder of the plant consists of different cells with anticlinal walls and diacytic stomata. Stomata have two subsidiary cells and the mentioned cells have a common wall at right angles to the longitudinal axis of the guard cells (Fig. 3.13).

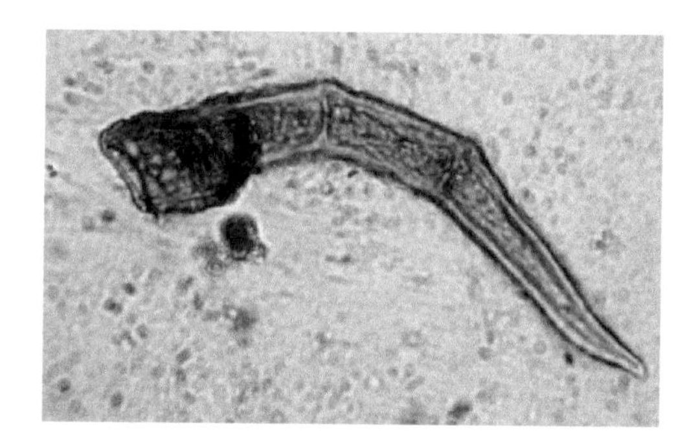

Fig. 3.12 A uniseriate covering trichome in the stem parts of *S. bachtiarica*

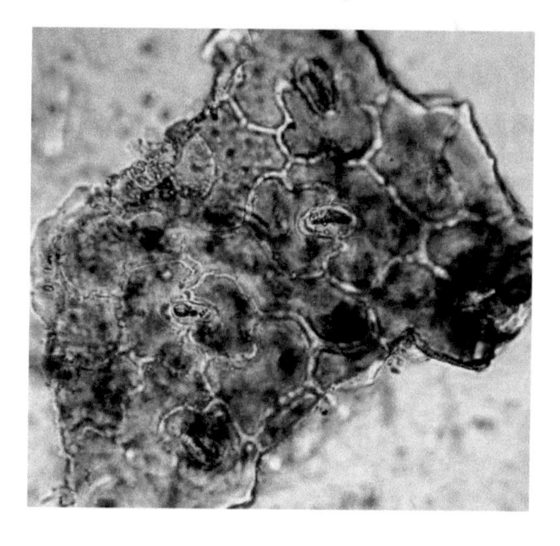

Fig. 3.13 Epidermis layer of stem in *S. bachtiarica* in the surface view, showing diacytic stomata

3.3 *S. hortensis* L.

3.3.1 Leaf

The epidermis of the leaf consists of cells with sinus walls and diacytic stomata (Fig. 3.14). Stomata in the leaf of *S. hortensis* have two subsidiary cells and these subsidiary cells possess a common wall at right angles to the longitudinal axis of the guard cells.

The uniseriate covering trichome with acute (or round tip) and slightly thickened cell wall are abundant in upper and lower epidermis of the leaf sections of *S. hortensis* (Fig. 3.15a, b, c, d). Some of the covering trichomes contain just one cell (Fig. 3.15c).

Glandular trichome with unicellular glandular scale represents a short stalk (Fig. 3.16). This type of glandular trichome is often found in the members of Lamiaceae family and therefore can be helpful in identification of unknown plant material.

Another type of glandular trichome has also been discovered in the leaf segments of *S. hortensis* with multicellular scales consisting of a short stalk and 12 secretory cells (Fig. 3.17). The glandular trichome is placed in a cavity of leaf epidermis cell (Fig. 3.17a). The epidermal cells surrounding the glandular trichome arranged to form a rosette shape (Fig. 3.17b).

Fragments of annularly and spirally thickened vessels attached to the fibers have been identified in the powder of *S. hortensis* leaf (Fig. 3.18).

Furthermore, some cluster crystals of calcium oxalate have been recognized near the vessels (Fig. 3.19). This type of crystal is made of the spheroidal aggregates of calcium oxalate with numerous faces and sharp points.

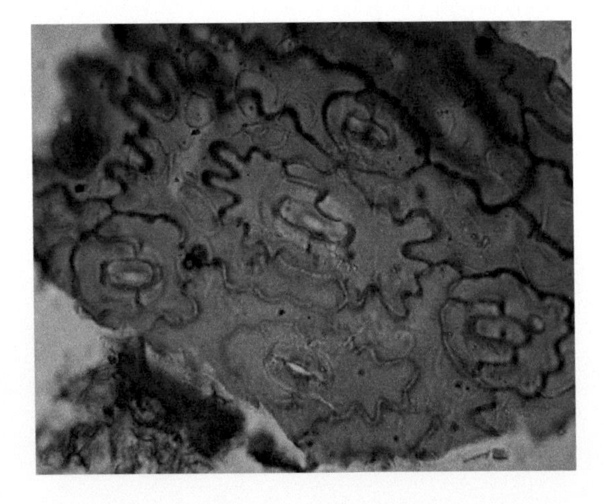

Fig. 3.14 The epidermis in the leaf powder of *S. hortensis* indicating diacytic stomata

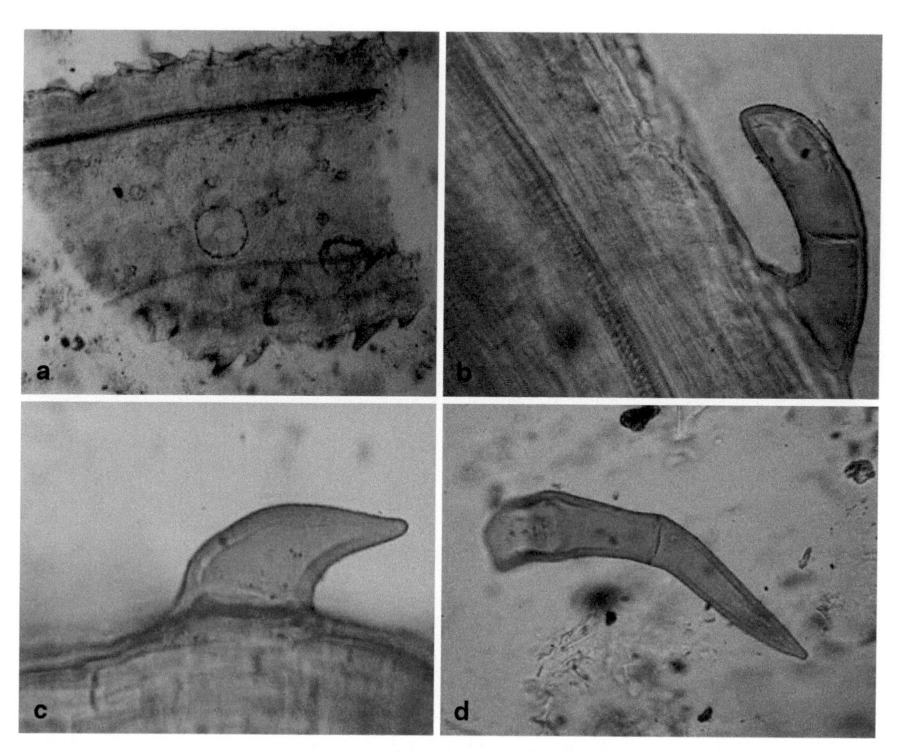

Fig. 3.15 Different types of covering trichomes in the leaf section of *S. hortensis*

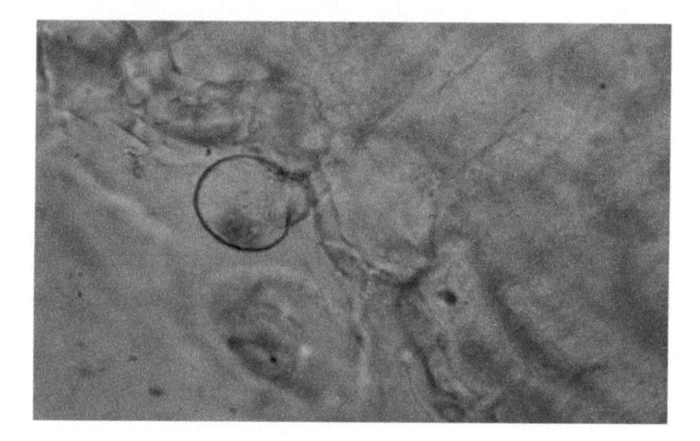

Fig. 3.16 A glandular trichome of *S. hortensis* with unicellular scale and a short stalk

3.3.2 Stem

Covering trichomes have been also scattered in the stem fragments of *S. hortensis*. They are uniseriate conical shape trichomes with several cells and acute or round tips. The trichomes sometimes curved at the apex. Commonly, the walls of the cells in the covering trichomes are slightly thickened and the cells contain some substances (Fig. 3.20).

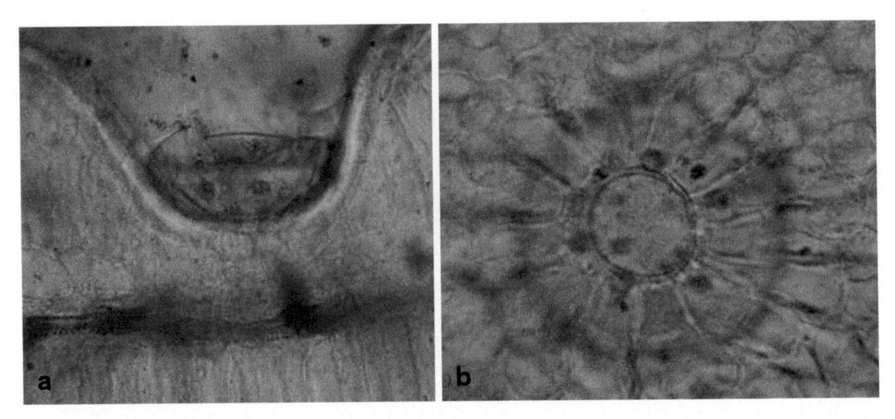

Fig. 3.17 Glandular trichomes in the leaf segments of *S. hortensis*; **a** in a cavity of epidermis cell; **b** arranged to rosette

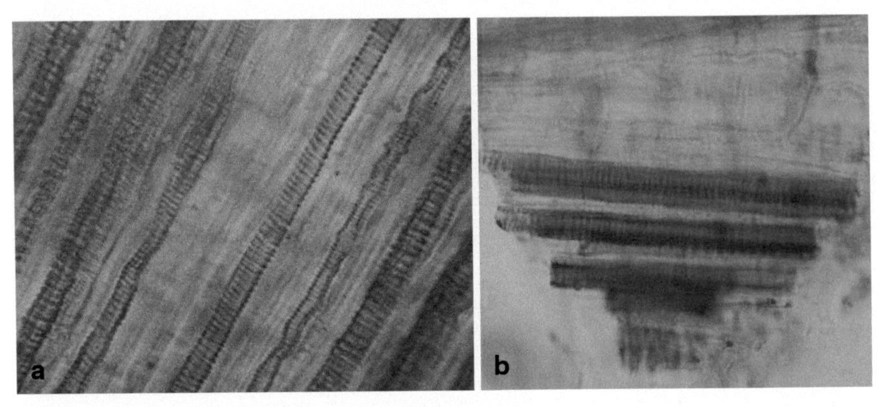

Fig. 3.18 Strands of annularly and spirally thickened vessels in the leaf powder of *S. hortensis*

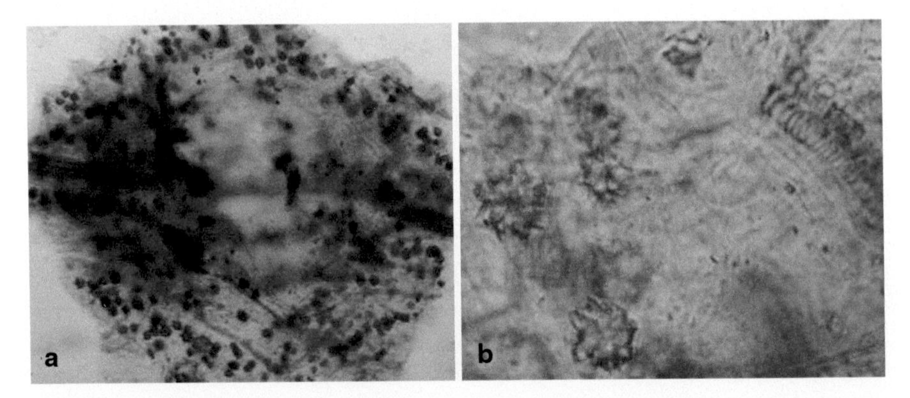

Fig. 3.19 Cluster crystal of calcium oxalate in the leaf powder of *S. hortensis*

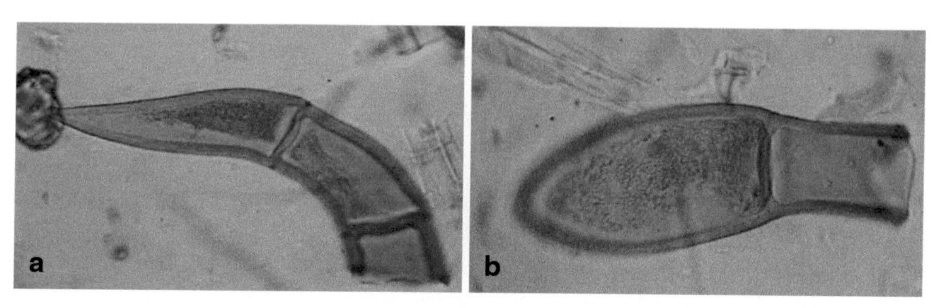

Fig. 3.20 Covering trichomes in the stem parts of *S. hortensis*

On the other side, the vascular tissues found in the stem segments of *S. hortensis* were single or in a large piece with fibers (Fig. 3.21). The fragments in the bordered pitted vessels were also determined in the stems of the plant (Fig. 3.21a). Vessels are lignified with spiral or annular thickening (Fig. 3.21b, c, d).

As the microscopic observations reveal, the epidermis cells in the stem powder of the plant compose of thin-walled and longitudinally elongated cells with occasional stomata, and also the glandular trichomes are similar to those discovered in the leaf of *S. hortensis* (Fig. 3.22).

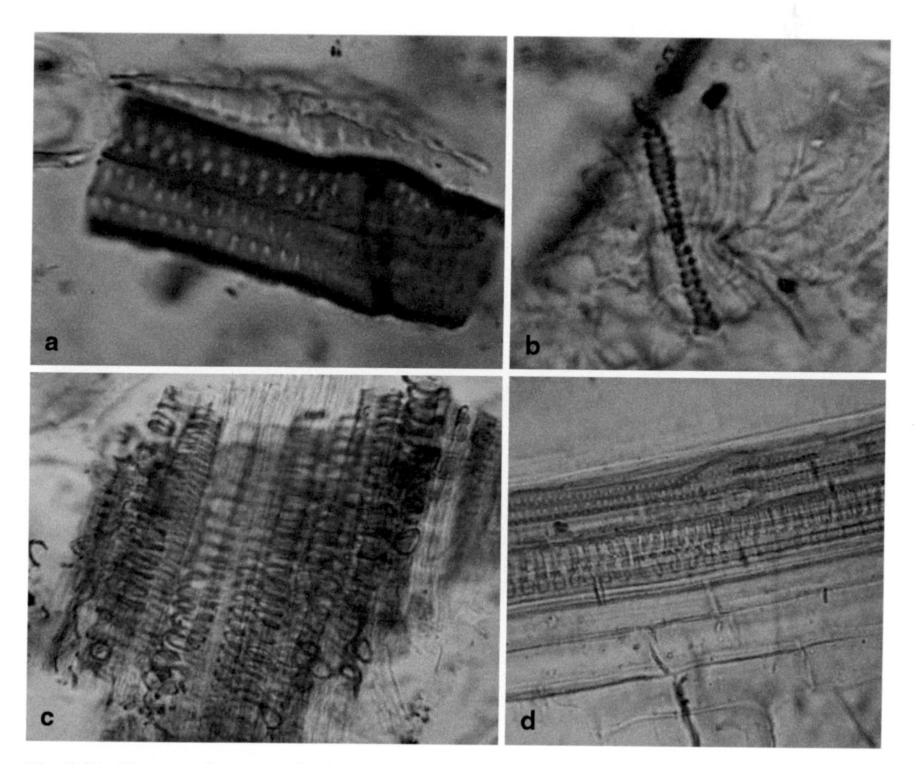

Fig. 3.21 The vascular tissues in the stem fragments of *S. hortensis*; (**a**) fragments in the bordered pitted vessels; (**b**) spiral; (**c**) and (**d**) annular thickening

3.4 *S. atropatana* Bunge

3.4.1 Leaf

The epidermis in the leaf segments of *S. atropatana* consists of the cells with anticlinal walls and diacytic stomata. Generally, stomata have two subsidiary cells, and these subsidiary cells possess a common wall at right angles to the longitudinal axis of the guard cells (Fig. 3.23).

Branching non-glandular trichome with acute tips and thickened cell walls has been found in the leaf segments of *S. atropatana* (Fig. 3.24).

Non-glandular covering trichome in the leaf segments of *S. atropatana* consists of several cells with slightly thickened cell walls. The terminal cell is in conical shape. The trichome become orange-red when treated with Sudan red that shows the presence of essential oil in the cells (Fig. 3.25).

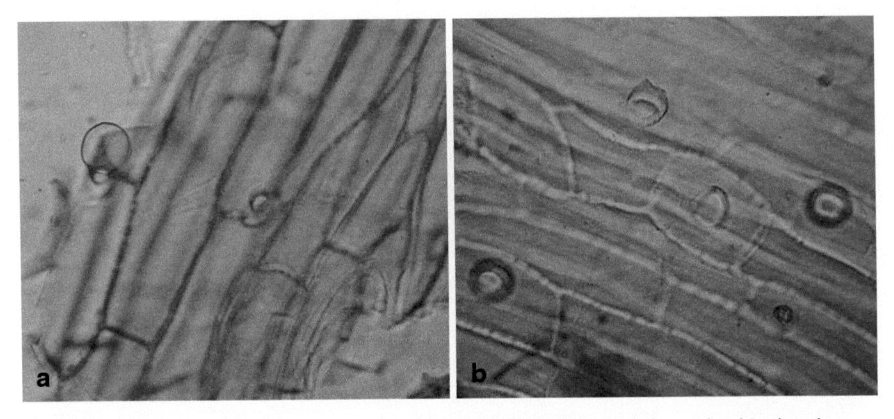

Fig. 3.22 Epidermis of the stem in *S. hortensis* in a surface view; **a** capitate gland **b** cicatrix

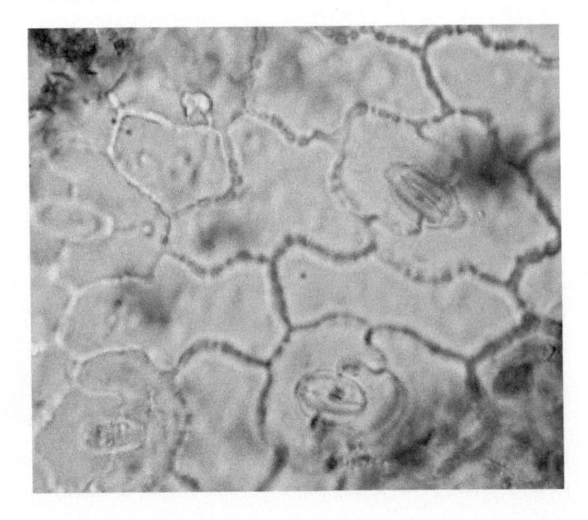

Fig. 3.23 Epidermal cells in the leaf segments of *S. atropatana* showing anomocytic stomata

Fig. 3.24 Non-glandular covering trichome in the leaf powder of *S. atropatana*

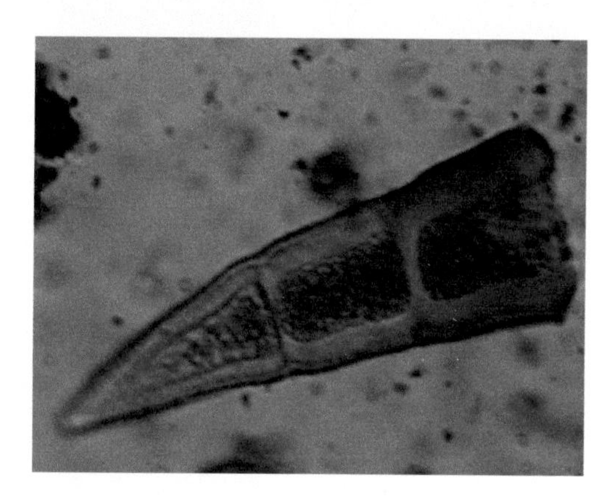

Fig. 3.25 Covering trichome in the leaf segments of *S. atropatana*

3.4.2 Stem

The microscopic observation shows that the parenchyma possesses thin wall in the stem powder of *S. atropatana*. The cells are fairly large and elongated to arrange with their long axes parallel to another (Fig. 3.26).

Fragments of the bordered pitted vessels in a surface view have been determined in the stem segments of *S. atropatana* (Fig. 3.27).

The sclereids fragments have been found in the stem of *S. atropana*. They are very large with oval shape thickened, pitted cell walls and large lumen with blunt ended (Fig. 3.28).

Moreover, uniseriate covering trichomes (known as non-glandular trichomes) with an acute or round tips and warty coticule are abundant in the stem segments of *S. atropatana* very similar to those found in the leaf segments of this species. They have several cells with thickened cell walls and they are usually fractured with reddish-brown content. This kind of trichome stained orange when the leaf powder treated with Saudan red, which indicates that the covering trichomes should contain the volatile oil (Fig. 3.29).

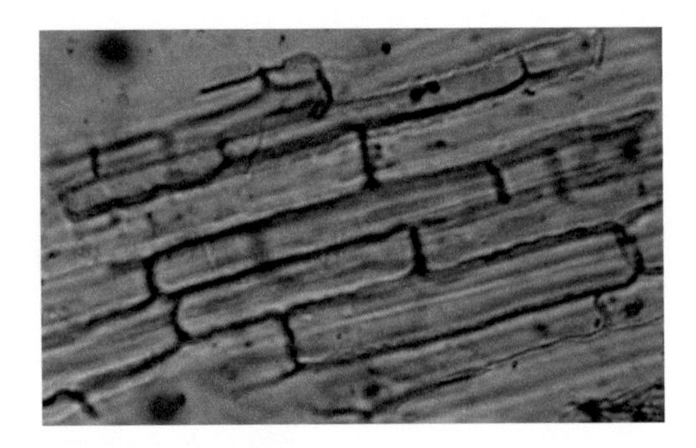

Fig. 3.26 Parenchyma in the stem fragments of *S. atropatana*

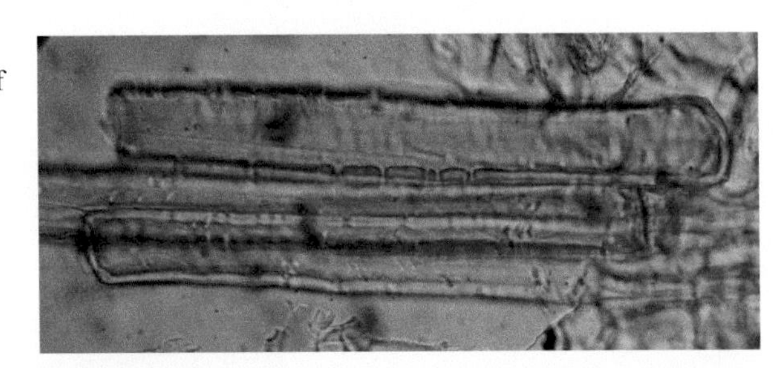

Fig. 3.27 Bordered pitted vessels in the stem powder of *S. atropatana*

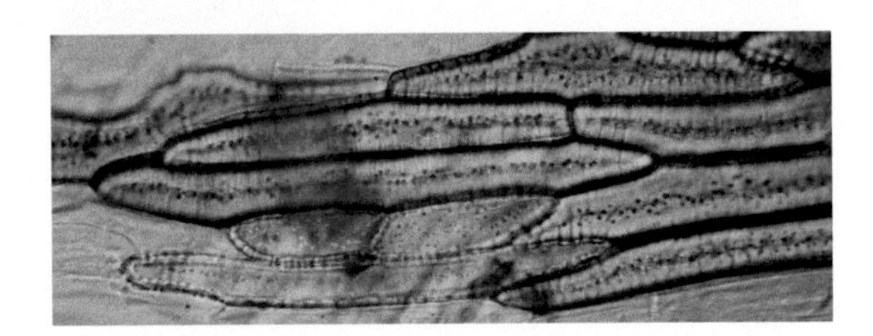

Fig. 3.28 Fibrous sclereids found in the stem fragments of *S. atroptaana*

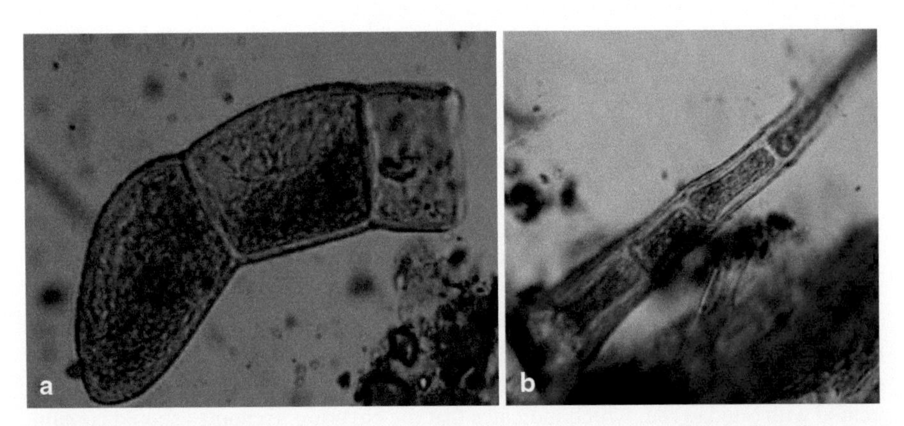

Fig. 3.29 Covering trichomes in the stem powder of *S. atropatana*; **a** round tip **b** sharp tip

3.5 *S. macrantha* C. A. Mey.

3.5.1 Leaf

The cells in the epidermis of the leaf parts of *S. macrantha* are observed as skew polygonal with dotted thin-cell walls. Actually, stomata are exhibited in a diacytic conformation, in that one of the subsidiary cells is bigger than the other one just like other Lamiaceae plants (Fig. 3.30).

The glandular trichomes in the leaf powder of *S. macrantha* are abundantly found in two different types similar to other *Satureja* species (Fig. 3.31a). In the first type, the larger one has a short stalk with a glandular head that has several cells with a common cuticle raised to from a spherical, bladder like trichome (Fig. 3.31b, d). The epidermal cells surrounding this glandular trichome generally arranged to form a rosette shape (Fig. 3.31c). The cells in the second type are smaller and capitate with a unicellular stalk and a rounded head composed of just one cell (Fig. 3.31b).

The calcium oxalate crystals are abundantly found in the leaf parts of *S. macrantha*. They are located in parenchymatous cells with cluster shape near vessel bundles (Fig. 3.32).

3.5.2 Flower

Microscopic observation reveals that the floral parts of *S. macrantha* represent fibrous layer of anther in a surface view (Fig. 3.33a) and also in a sectional view (Fig. 3.33b). This structure is useful for identification of a material of floral origin not for identification of the different species of *Satureja*, since it is commonly observed in various species.

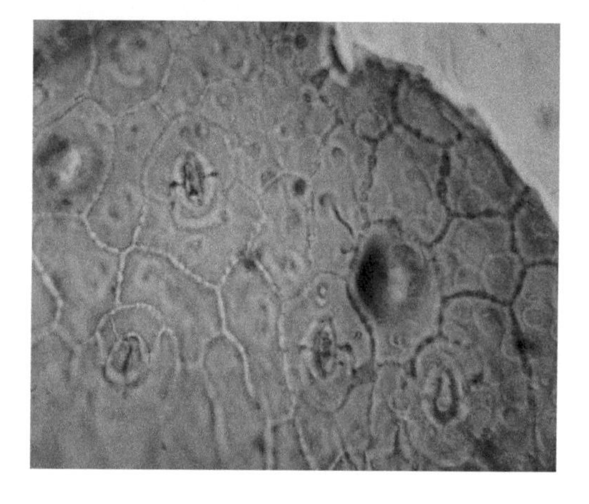

Fig. 3.30 An epidermis view of the leaf parts of *S. macrantha* showing diacytic stomata

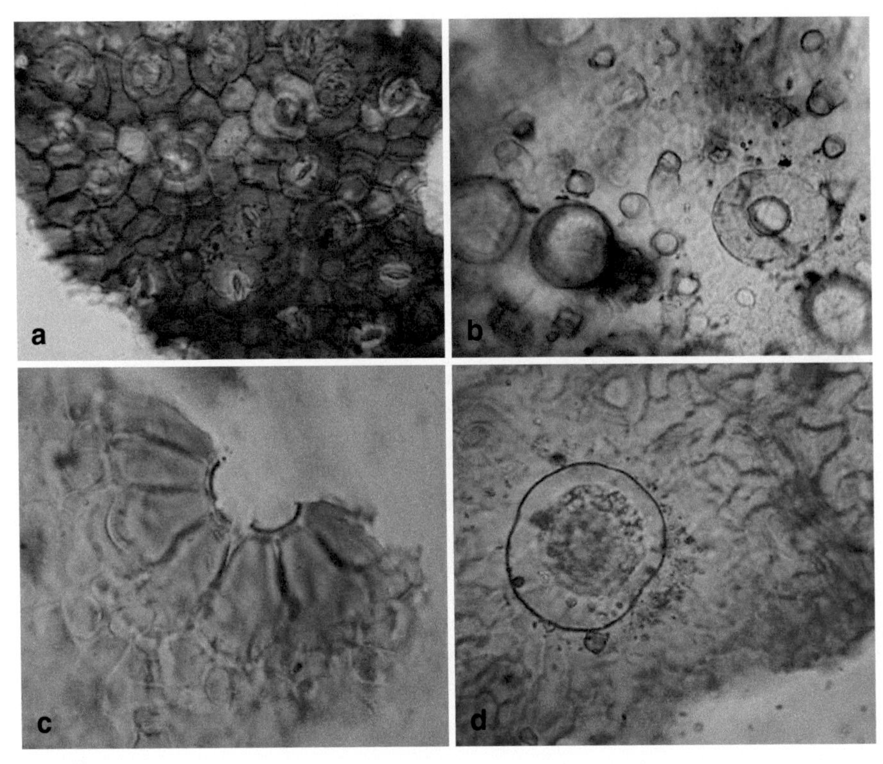

Fig. 3.31 Epidermis cells in the leaf segments of *S. macrantha*; **a** diacytic stomata and glandular trichomes; **b** capitate and multicellular glands; **c** radiating epidermal cells showing rosette; **d** a glandular gland

Fig. 3.32 Crystals of calcium oxalate in the leaf parts of *S. macrantha*

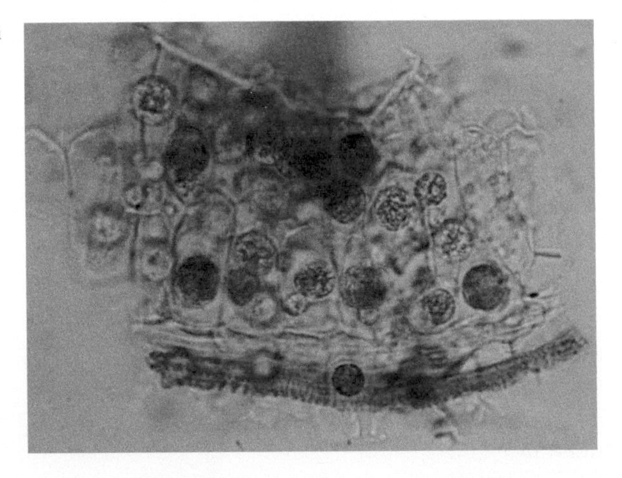

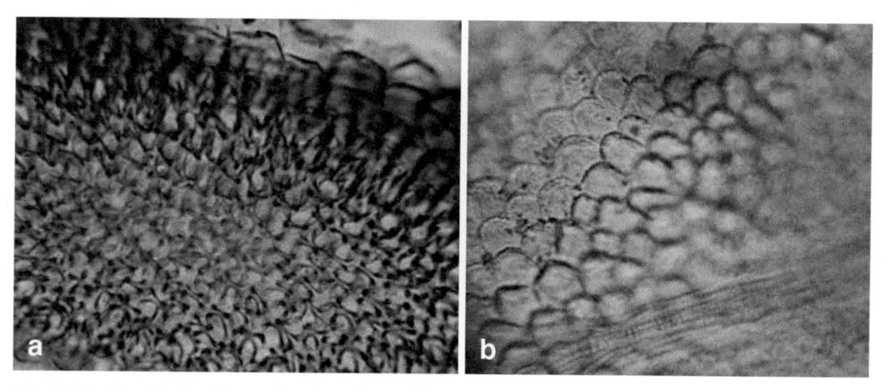

Fig. 3.33 The fibrous layer of anther in the floral segments of *S. macrantha*; **a** in a surface view; **b** in a sectional view

Fig. 3.34 Calyx of *S. macrantha* exhibiting some trichomes

The covering trichomes have abundantly been found in the calyx segments of *S. macrantha* particularly on the outer epidermis and on the lobe. The trichomes are similar to those found in the leaf parts of the plant containing reddish-brown substances. Glandular trichomes are fairly existed particularly in capitate type (Fig. 3.34)

3.5.3 Stem

The epidermis of stem in *S. macrantha* composes of elongated cells arranged with their long axes along each other in the surface view. Cicatrix is a characteristic scar of cleavage in a trichome that can be observed in the stem parts of the plant (Fig. 3.35).

Both covering trichomes and glandular trichomes can be found in the stem parts of the plant. The glandular trichome has stalk consisting of two cells and oval shape head with reddish-brown content (Fig. 3.36a). The covering trichome has several

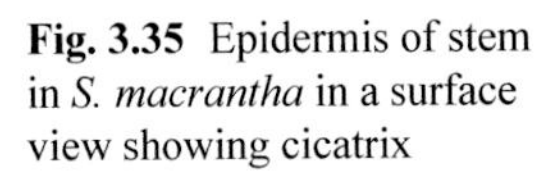

Fig. 3.35 Epidermis of stem in *S. macrantha* in a surface view showing cicatrix

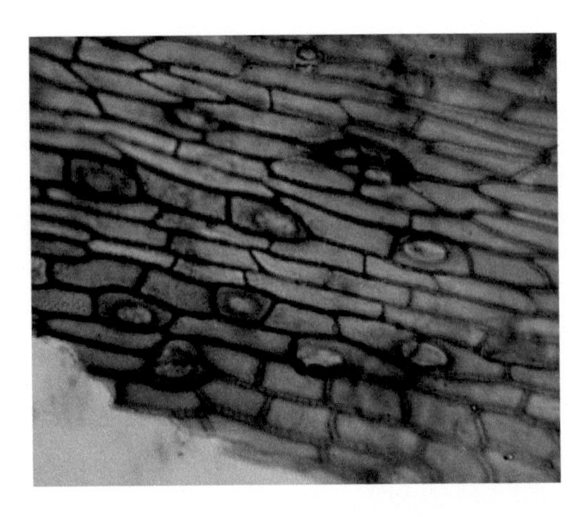

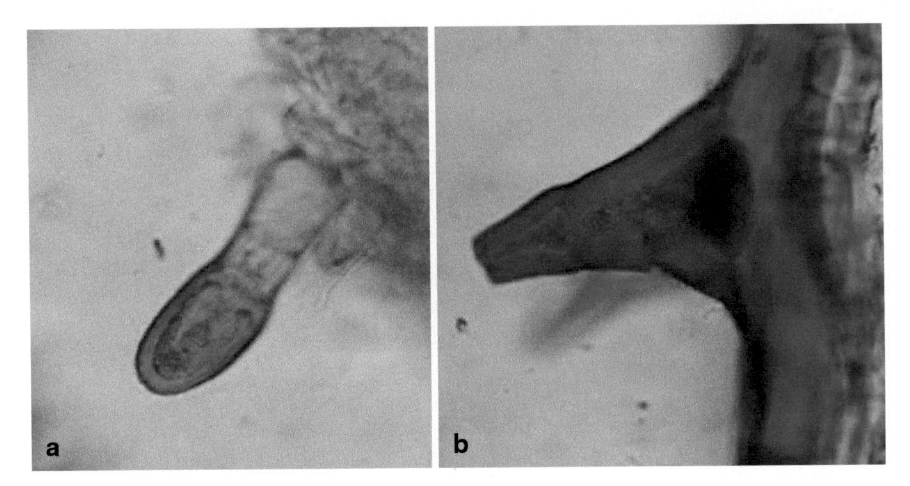

Fig. 3.36 Trichomes of stem in *S. macrantha* in the surface view

cells with thickened cell walls in the surface view, and also it seems to be filled with dense contents (Fig. 3.36b).

The fragments of cork compose of cells with slightly thickened cell walls in a surface view. The cells are polygonal in different size (Fig. 3.37).

The vessels in the stem powder of *S. macrantha* are found single or in the large groups. The walls are lignified with spiral or annular thickenings (Fig. 3.38a, b). Associated fibers are observed too that composed of elongated cells with thin walls (Fig. 3.38c).

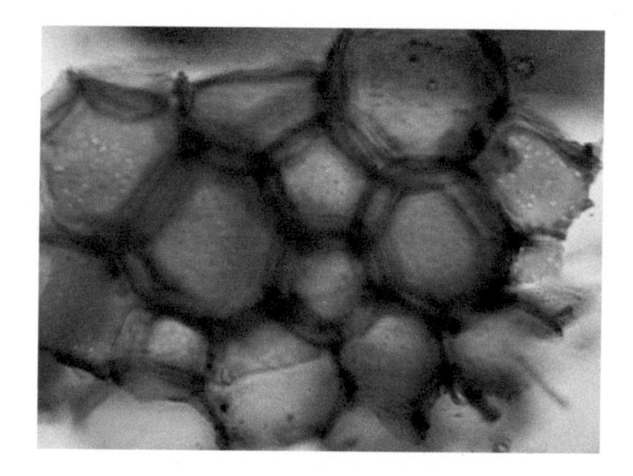

Fig. 3.37 Cork in the stem parts of *S. macrantha* in a surface view

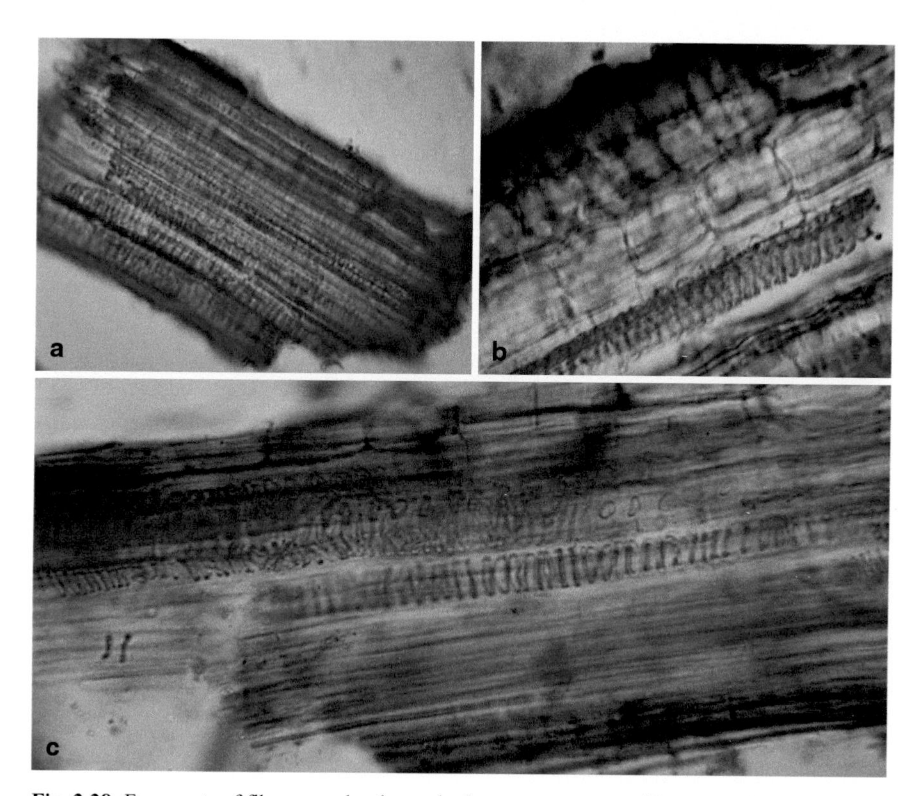

Fig. 3.38 Fragments of fibro-vascular tissues in the stem segments of *S. macrantha*

Chapter 4
Phytochemical Contents

4.1 Rosmarinic Acid

Different concentrations of rosmarinic acid, as a natural phenolic compound, have been reported in the various extracts of Lamiaceae family (Nepetoideae subfamily) ranging between 0.001 and 0.93 % [16]. In a study, 30 samples *S. hortensis*, collected from different parts of Iran, were analyzed using high performance thin layer chromatography (HPTLC) and the results indicated a considerable variation in the content of this compound ranging from 0.06 to 0.69 % based on dry weight [17]. However, in some other studies, the concentrations of rosmarinic acid were evaluated as 2.5 % in the ethanol extract of *S. hortensis* and 1.2 % in the aerial parts of the plant as well [18, 19]. In addition, the rosmarinic acid content in the ethanol and acetone extracts of *S. hortensis* were analyzed using nucleic magnetic resonance (NMR) and high performance liquid chromatography (HPLC) methods resulted in a higher measured amount of rosmarinic acid calculated in the ethanol extract rather than the acetone extract [20]. Furthermore, the content of rosmarinic acid was determined in some Iranian species including *S. atropatana*, *S. bachtiarica*, *S. hortensis*, *S. khuzistanica*, *S. macrantha*, and *S. mutica* using HPLC method, and the amount of the compound in the above mentioned plants were reported 2.8, 5.7, 16.3, 1.2, 4.2, and 19 mg/g, respectively [21]. In compared with the previous data, all the tested species contained different rosmarinic contents (0.001–2.5 %). The mentioned analyzing methods (HPTLC, HPLC and NMR) are generally employed for determination and phyto-analysis of various secondary metabolites in the plants as well.

4.2 Phenolic Compounds and Flavonoids

A literature review demonstrated that the phenolic compounds as well as flavonoids have commonly been reported from *Satureja* species. For instance, the percentage of phenolic compounds was calculated in an aqueous extract of *S. montana* using

S. Saeidnia et al., *Satureja: Ethnomedicine, Phytochemical Diversity and Pharmacological Activities*, SpringerBriefs in Pharmacology and Toxicology,
DOI 10.1007/978-3-319-25026-7_4

HPLC method followed by MS/MS. The results showed that the extract was reach in caffeic acid (76.94 %) as well as rutin (20.36 %). Other phenolic compounds such as chlorogenic acid, naringenin, coumaric aicd, and protocatechuic acid were identified in the extract of winter savory in lesser amount [22]. The ethanol extract of *S. hortensis* also contained a trace amount of caffeic acid but this compound was not identified in acetone extract of the plant. In the acetone extract of summer savory, two flavones luteolin and apigenin were detected by observation of −OH [5] resonance in 12–13 ppm in ^{1}H-NMR spectra [20]. In addition, caffeic acid, luteolin-glucoside, naringenin-glucoside, apigenin-glucoside, rosmarinic acid, eriodictyol, and apigenin were detected in the extract of *S. hortensis* and qualitatively analyzed using HPLC with Photo Diode Array (PDA) detector. An increase in total phenol content of the mentioned extract was significantly observed by acid treatment [23].

HPLC analysis of *S. hortensis* confirmed the presence of a number of phenolic acids (caffeic and *p*-coumaric acids), flavonoid aglycones (catechin, epicatechin, luteolin and apigenin), and flavonoid glycosides (rutin, hesperidin, apigenin-7glucoside) in its ethanol extract [24]. The main compounds in the aerial parts of *S. hortensis* were identified as flavonoids like apigenin and apigenin-4′-methyl ether [25, 26]. Presence of 5,6,4′-trihydroxy-7,3′-dimethoxyflavone and 5,6-dihydroxy-7,3′,4′-trimethoxyflavone were assessed in *S. thymbra* as well as 5,6,4′-trihydroxy-7,8,3′-trimethoxyflavone (thymonin), 5,6,4′-trihydroxy-7,8-dimethoxyflavone (thymusin) in *S. salzmannii* [27]. Moreover, acacetin 7-O-rhamnosyl [1‴→6″] glucoside was detected in aqueous methanol (80 %) extracts of *S. kitaibelii*, *S. cuneifolia*, *S. montana* spp. *montana*, and *S. montana* spp. *variegate* using HPLC [28]. In that study, the main flavonoids of *S. thymbra* and *S. spinosa* were characterized and identified using retention time, UV and MS spectrums in comparison with the reported standard data. Genkwanin, naringenin, aromadendrin, eriodictyol, taxifolin, and 6-hydroxyluteolin 7,3′-dimethyl ether were finally reported in both species. Apigenin, xanthomicrol, cirsimaritin, thymusin, thymonin, cirsilineol, and 8-methoxycirsilineol were also identified only in *S. spinosa*, whereas luteolin 7-methyl ether, ladanein, and 6-hydroxyluteolin 7,3′,4′-trimethyl ether were found only in *S. thymbra* [29]. Additionally, phytochemical investigation of *S. acinos* resulted in purification of naringenin, rutinoside, and ery(i)odictyol [30]. Moreover, four flavonoids were isolated from *S. atropatana* named 5,6,3′-trihydroxy-7,8,4′-trimethoxyflavone, 5,6-dihydroxy-7,8,3′,4′-tetramethoxyflavone (5-desmethoxynobiletin), 5,6,4′-trihydroxy-7,8,3′-trimethoxyflavone (thymonin) and luteolin (Fig. 4.1) [31].

4.3 Triterpenes

Studies exhibited that different triterpenes in the ethyl acetate extract of winter savory were quantitatively analyzed using gas chromatography-flame ionization detector (GC-FID) and high performance liquid chromatography-photodiode array (HPLC-PDA) apparatus. The amounts of betulinic acid (BA), oleanolic acid (OA),

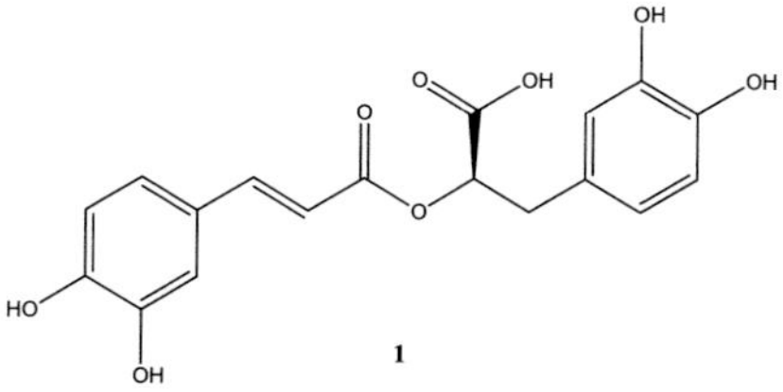

1

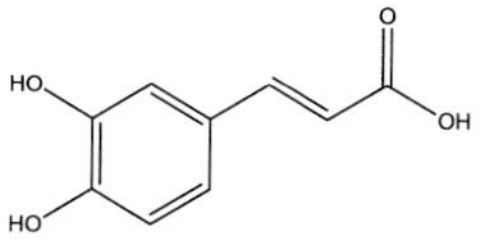

2

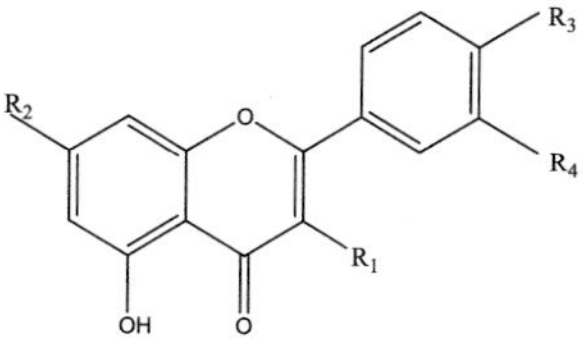

3 R_1=O-rutinoside R_2=OH R_3=OH R_4=OH
22 R_1=H R_2=O-rutinoside R_3=OCH_3 R_4=H
23 R_1=H R_2=OCH_3 R_3=OH R_4=H

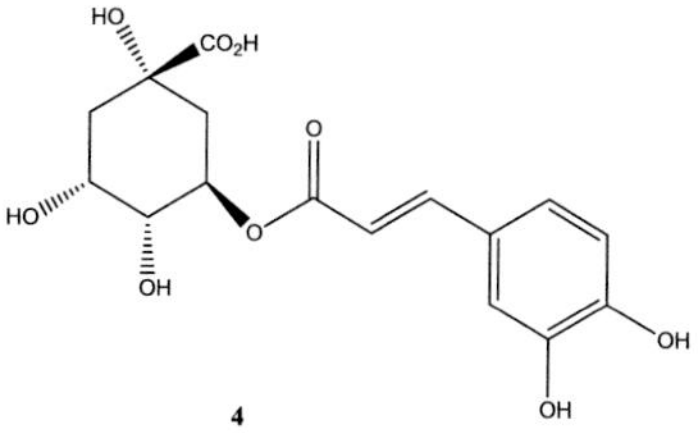

4

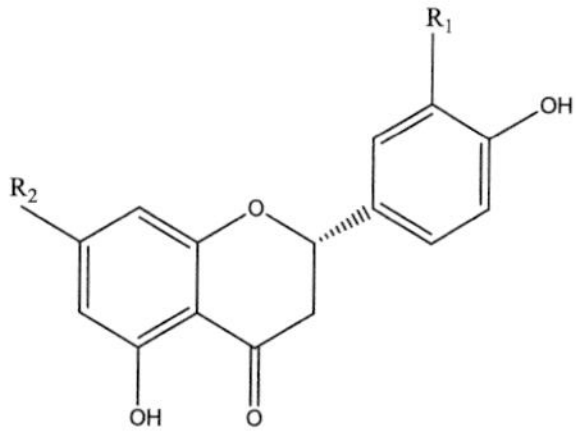

5 R_1=H R_2=OH
11 R_1=H R_2=O-glucoside
13 R_1=OH R_2=OH

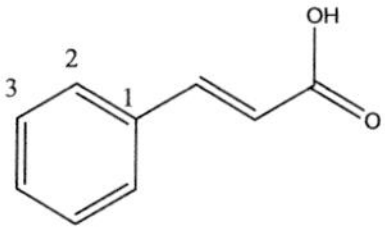

6 1- ortho OH
2- Meta OH
3- Para OH

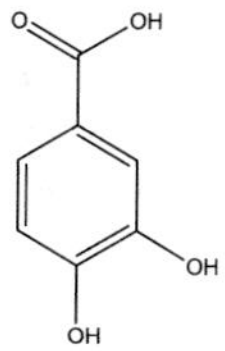

7

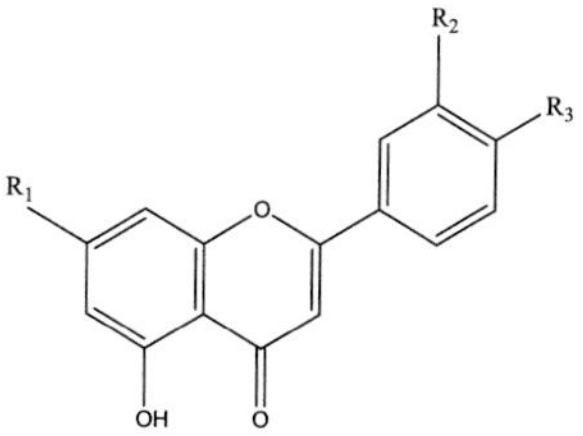

8 R_1=OH R_2=OH R_2=OH
9 R_1=OH R_2=H R_3=OH
10 R_1=O-glucoside R_2=OH R_3=OH
12 R_1=O-glucoside R_2=H R_3=OH
17 R_1=OH R_2=H R_3=OCH_3
31 R_1=OCH_3 R_2=OH R_3=OH

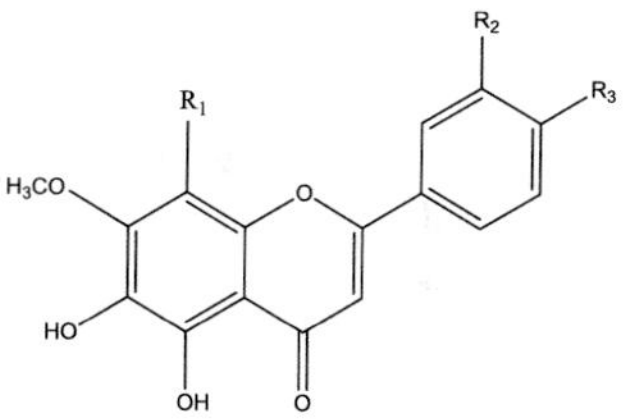

18 R_1= H R_2=OCH_3 R_3=OH
19 R_1= H R_2=OCH_3 R_3=OCH_3
20 R_1= OCH_3 R_2=OCH_3 R_3=OH
21 R_1= OCH_3 R_2=H R_3=OH
34 R_1= OCH_3 R_2=OH R_3=OCH_3
35 R_1= OCH_3 R_2=OCH_3 R_3=OCH_3

Fig. 4.1 Chemical structures of the isolated compounds from various species of *Satureja*

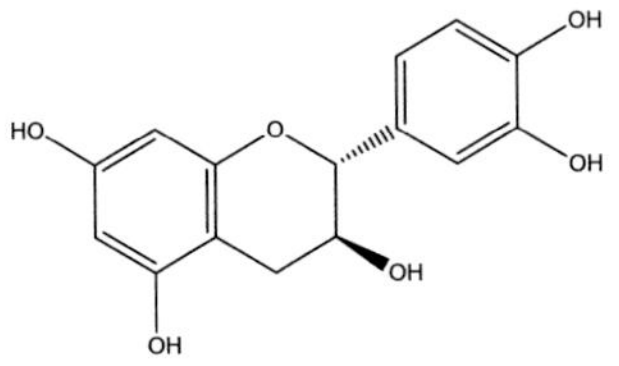

14

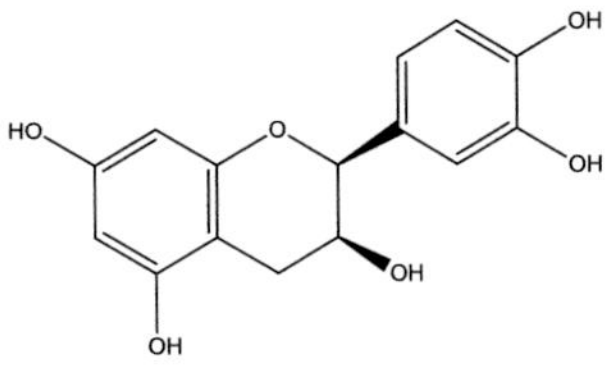

15

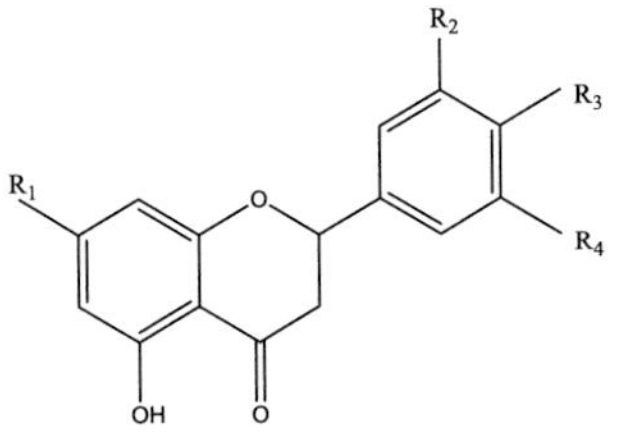

16 R_1= O-rutinoside R_2=OH R_3= OCH_3 R_4=H
36 R_1= OH R_2= OH R_3= H R_4=OH
37 R_1=(6"-O-alpha-L-rhamnopyranosyl)-beta-D-glucopyranoside R_2=OCH_3 R_3= OH R_4=H

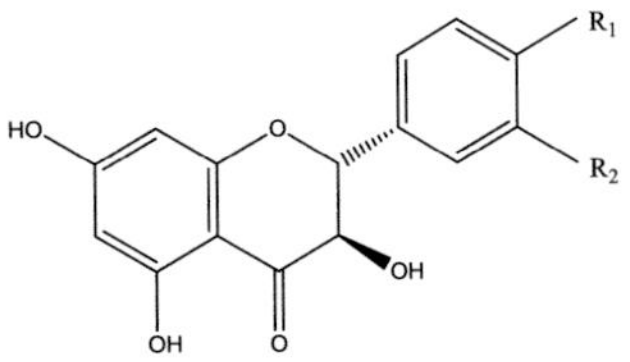

24 R_1= OH R_2=H
25 R_1= OH R_2=OH

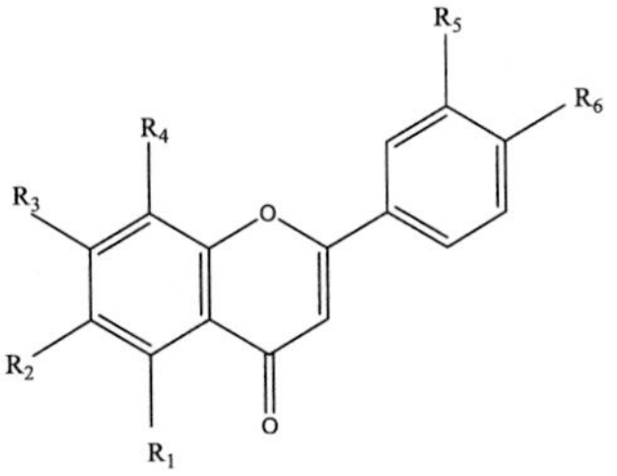

26 R_1=OH R_2=OH R_3= OCH_3 R_4=H R_5= OCH_3 R_6=OH
27 R_1=OH R_2=OCH_3 R_3= OCH_3 R_4=OCH_3 R_5= H R_6=OH
28 R_1=OH R_2=OCH_3 R_3= OCH_3 R_4=H R_5= H_3 R_6=OH
29 R_1=OH R_2=OCH_3 R_3= OCH_3 R_4=H R_5= H R_6=OH
30 R_1=OH R_2=OCH_3 R_3= OCH_3 R_4=H R_5= OCH_3 R_6=OH
32 R_1=OH R_2=OH R_3= OCH_3 R_4=H R_5= H R_6=OCH_3
33 R_1=OH R_2=OH R_3= OCH_3 R_4=H R_5= OCH_3 R_6=OCH_3

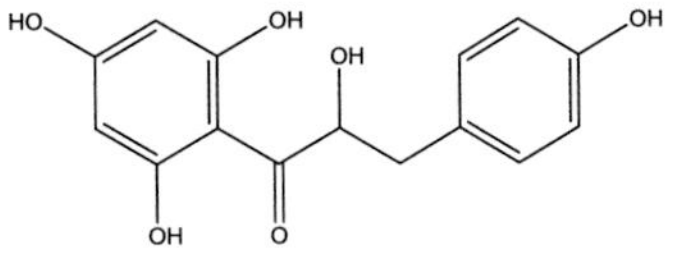

38

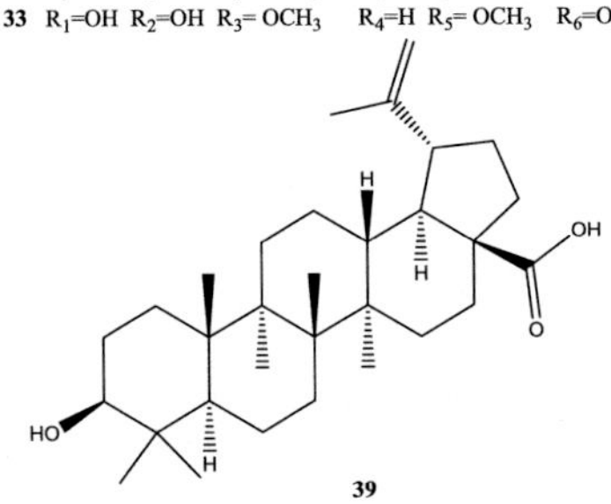

39

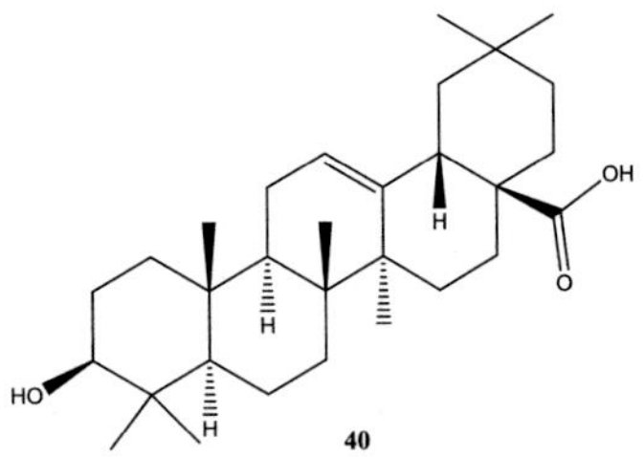

40

Fig. 4.1 (continued)

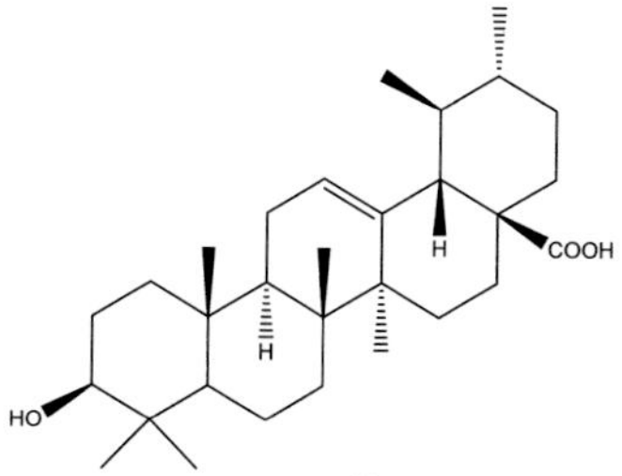

41

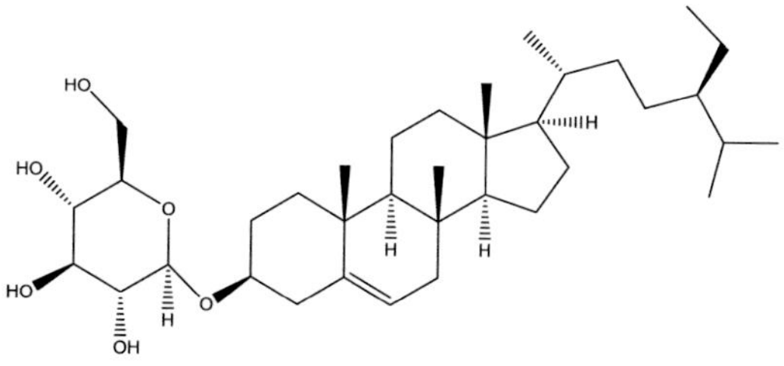

42

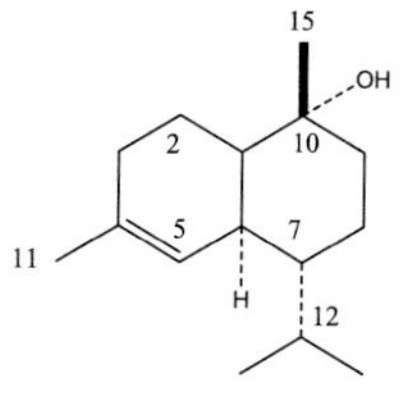

43

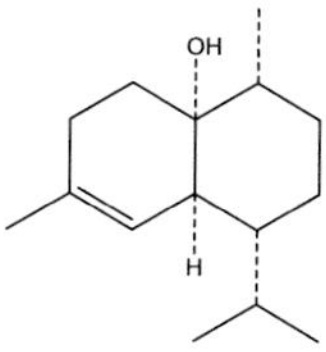

44

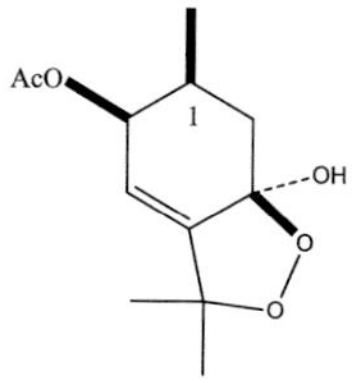

45

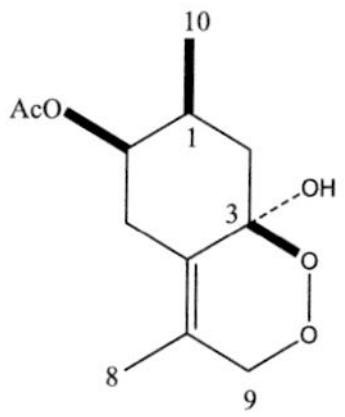

46

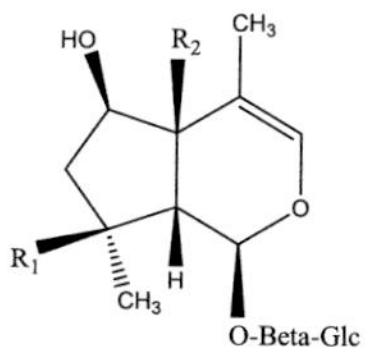

47 R_1= OH R_2=H
49 R_1= OH R_2=OH
50 R_1= $OCOCH_3$ R_2=H

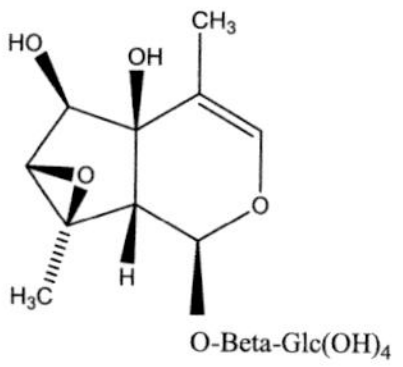

48

Fig. 4.1 (continued)

and ursolic acid (UA) in the extract were evaluated as 0.04, 0.14, and 0.49 %, respectively [32]. In another study, the amount of the mentioned compounds in ethanol extract of the plant were measured as 0.043, 0.536 and 0.094 %, respectively [33]. Fuethermore, the methanolic extracts of *S. montana* and *S. coeruela* were obtained by soxhlet apparatus and analyzed with GC-FID method. The amount of OA in *S. montana* and *S. coeruela* extracts were calculated as 0.131 and 0.176 %, while for UA, the values were 0.490 and 0.644 %, respectively [34]. The amount of OA in *S. mutica* was calculated as 17.5 mg per 100 g dried leaves of the plant using densitometric analysis of the developed plate of TLC suggesting the plant as an industrial source of OA [35]. Chromatography of the acetone extract of *S. acinos* resulted in purification of OA, UA, and daucosterol as well [30].

4.4 Other Classes of Secondary Metabolites

Bioassay guided isolation of dichloromethane extract of *S. gilliesii* by brine shrimp test resulted in isolation and identification of two sesquiterpene alcohols with cadinane skeleton named (+)-T-cadinol and (−)-cadin-4-en-1-o1 along with two monoterpenes with menthane-derived bicyclic skeleton named acetylsaturejol and isoacetylsaturejol [36]. Four iridoid glycoside (5-deoxylamiol, 4-methylantlrrhinoside, lamiol and 5-deoxylamloslde) were purified from flowering parts of *S. vulgaris* [37].

4.4.1 Essential Oil

Essential oil composition of some *Satureja* species revealed three main chemotypes classified as:

I) Aromatic *p*-menthane monoterpenes (chemotype I: mostly carvacrol, thymol and *p*-cymene).
II) Aliphatic *p*-menthane monoterpenes (chemotype II: mostly menthone, isomenthone, pulegon, and piperitone).
III) Several mono- and sesquiterpenes (chemotype III) (Table 4.1) [38].

Different drying and extraction methods followed by genetic, phenological stage of plant development, day light intensity, temperature and climate showed noticeable effect on the yield of essential oils and their constituents on *Satureja* species [39–43]. For instance, the results of a study revealed that different drying methods (sunshine, shade, and oven drying at 45 °C) along with various distillation methods (hydro-distillation, water- and steam-distillation, and steam distillation) affected the content and composition of the essential oil of *S. hortensis*. The major constituents in all the volatile oils were found as carvacrol and γ-terpinene, among them carvacrol (48.1 %) was the main compound in the oil of oven drying method, while γ-terpinene (70.4 %) was the main constituent of the oil by steam-distillation method [44].

Moreover, different procedures of extraction including supercritical fluid extraction (SFE) and hydro-distillation (HD) method resulted in different compositions of the respective oils. The color of SFE oil and HD oil were dark red and yellow, respectively that seems to be associated with the concentration of thymoquinone [45]. Subcritical water method was able to extract more polar (oxygenated) flavors in comparison with HD and SFE with pure CO_2. Actually, the latter method mainly resulted in extraction of alkane waxes and little amount of other flavors that might be due to the less solubility of the oxygenated compounds in pure CO_2 [46]. Differences were also found in composition of essential oil of *S. hortensis* in various parts of the plant including petals, calyces, young, medium and old leaves. Every oil gland was extracted by using SPME fiber and analyzed by GC-FID resulting in

Table 4.1 Major components of the essential oils of some *Satureja* species and their chemical structures

Plant name	Major compound	Structure	Chemotype	Reference
S. rechingeri, S. hortensis, S. khuzistanica, S. mutica, S. thymbra, S. montana ssp. *variegate, S. cuneifolia, S. subspicata, S. parnassica* ssp. *sipylea, S. parnassica* ssp. *parnassica, S. montana, S. icarica, S. pilosa, S. parvifolia, S. hortensis, S. boissieri, S. subspicata* subsp. *subspicata, S. cilicica, S. bachtiarica, S. horvatii* ssp. *macrophylla, S. spicigera, S. spinosa, S. cuneifolia*	Carvacrol	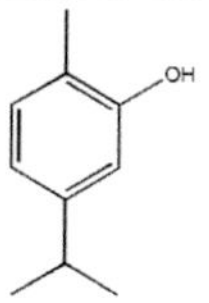	I	[12, 17, 43, 49, 55–77]
S. spicigera, S. cuneifolia, S. thymbra, S. pilosa var. *pilosa, S. intermedia, S. hortensis, S. mutica, S. atropatana, S. montana, S. bachtiarica, S. sahendica, S. horvatii* ssp. *macrophylla*	Thymol	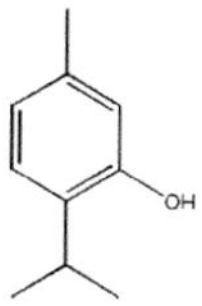	I	[12, 43, 57, 78–88]
S. aintabensis, S. khuzistanica, S. sahendica, S. macrantha, S. spicigera, S. edmondi, S. horvatii, S. montana, S. montana ssp. *pisidica, S. kitaibelii, S. obovata*	*P*-cymene	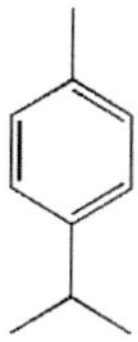	I	[39, 57, 67, 89–96]
S. darwinii	Piperitenone	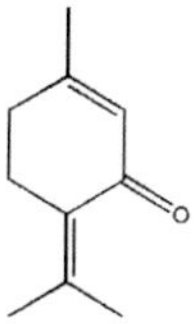	II	[38]
S. douglasii	Carvone	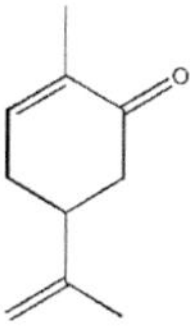	–	[42]
S. masukensis, S. pseudosimensis, S. parvifolia	Piperitenone oxide	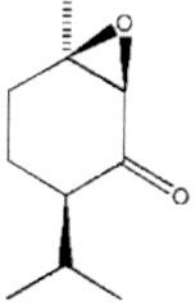	II	[97–100]

Table 4.1 (continued)

Plant name	Major compound	Structure	Chemotype	Reference
S. multiflora, S. glabella, S. boliviana, S. douglasii	Isomenthone	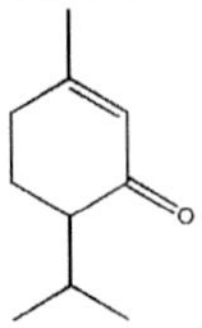	II	[5, 38, 98, 101, 102]
S. thymbra, S. boliviana	γ-terpinene	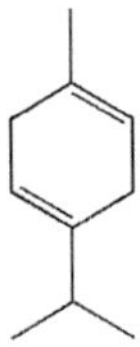	–	[99, 103]
S. brevicalix	Menthone	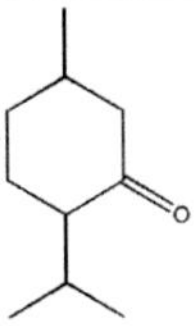	II	[5]
S. visianii, S. spinosa, S. cuneifolia, S. montana ssp. *montana, S. horvatii* ssp. *macrophylla*	Linalool	 OH	–	[43, 50, 104, 105]
S. punctata, S. forbesii	Geranial	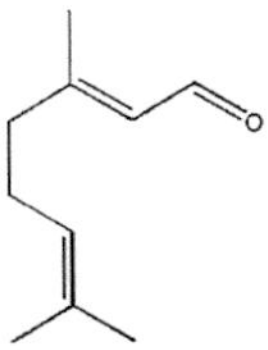	–	[106, 107]
S. odora, S. brownei	Pulegone	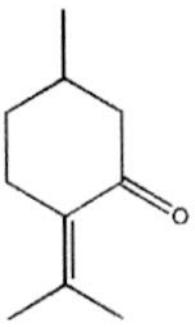	II	[64, 108]
S. fukarekii, S. adamovicii	α-phellandrene	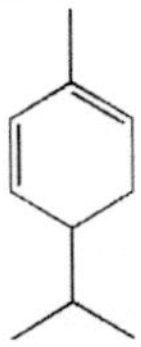	–	[109]

Table 4.1 (continued)

Plant name	Major compound	Structure	Chemotype	Reference
S. montana ssp. *kitaibelii*	Limonene	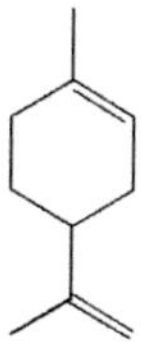	–	[81]
S. parnassica ssp. *parnassica*, *S. coeruela*	β-caryophyllene	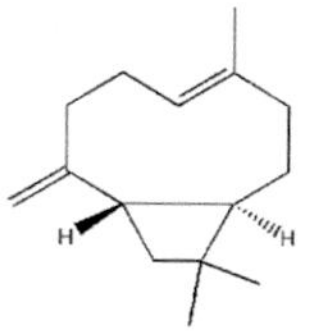	III	[77, 110]
S. glabrata	Isocaryophyllene	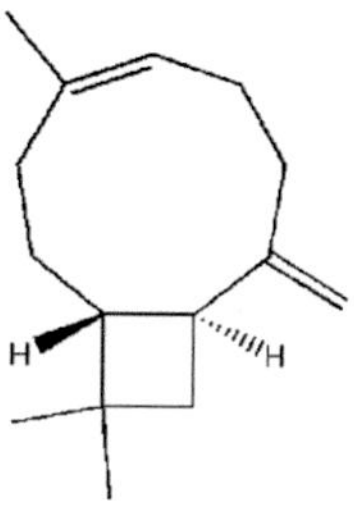	III	[111]
S. isophylla	α-eudesmol	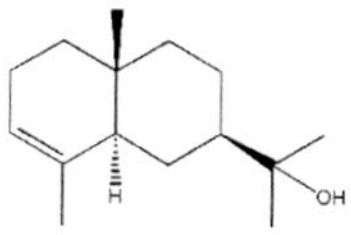	III	[93, 112]
S. macrantha, *S. biflora*, *S. parnassica* ssp. *parnassica*	Spathulenol	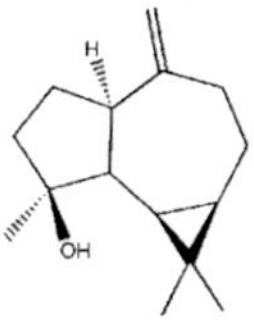	III	[83, 97, 110]
S. wiedemanniana	Caryophyllene oxide	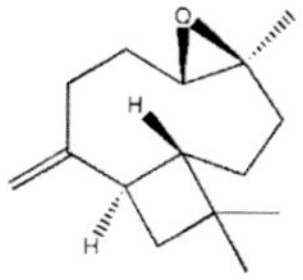	III	[113]
S. alpine, *S. coerulea*	Germacrene D	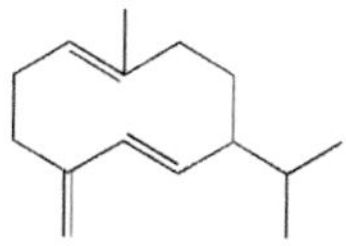	III	[63, 101, 114]

identification of carvacrol as a main component of the oils. Although carvacrol is the major compound of all the mentioned oils, γ-terpinene percentage in different age of leaf were higher than flower parts [47]. Additionally, the result of a previous study revealed that water stress treatment increased the amount of the essential oil of summer savory along with the constituents of the oils. The major compounds of both volatile oils were carvacrol and γ-terpinene. The content of carvacrol increased under moderate water stress, while the content of γ-terpinene decreased in moderate and sever water stress [48].

Moreover, the essential oil of *S. rechingeri* (in the beginning and full flowering stages) were analyzed with different distillation methods including steam, hydro- and water-steam-distillation. Their findings indicated that carvacrol was the major constituent of all the oils in different amounts, while the percentage of carvacrol increased at full flowering stage to 84.0–89.3 % [49]. Composition of the essential oils of *S. cuneifolia,* depending on the different stages of flowering period, was observed variable. Compounds linalool and borneol were relatively constant but carvacrol, limonene and α-pinene showed variability during the growth cycles [50]. In the earlier study, it is revealed that concentration of all components of *S. thymbera* oil varied periodically, and higher amount of thymol was measured in July [51]. Low water potential correlated with low monoterpenoid content in *S. douglasii* in field, which may attribute to reduction of photosynthesis, since there is correlation between monoterpenoid synthesis and photosynthesis [52]. Actually, the monoterpenoid yield per leaf weight in *S. douglasii* increased in low light regardless of day light. However, day temperature had a minor effect, while higher temperature resulted in slightly higher yield in the plant oil [41]. Thymol content of *S. obovata* oil increased simultaneously with the hardening of weather conditions [39].

The essential oil of *S. montana* was subjected to GC-MS analysis resulting in identification of thymol and *p*-cymene as the major oil components before hydrolysis, while after enzymatic hydrolysis using β-glycosidase, thymoquinone and eugenol presented as the main part of the volatile aglycones. It seems that enzymatic hydrolysis caused thymohydroquinone oxidizes to thymoquinone [53]. Both the composition and content of the essential oil obtained from *S. horvatii* ssp. *macrophylla* are found variable regarding to the altitude and geographical pattern. In Mediterranean and Sub Mediterranean bioclimates, carvacrol dominated, while in Sub Mediterranean and Temperate Axeric bioclimates, either linalool or *trans*-sabinene were found as the main part of the oil [43]. Literature review demonstrated that various essential oils in different *Satureja* species possess similar composition. Also, the plant habitat influences on the composition of the plant oil, even different species in the same habitat are able to produce similar composition [54]. Most of the *Satureja* species are categorized in chemotype I (Table 4.1), in which carvacrol, thymol and *p*-cymene were the main part of the essential oils. Mediterranean and Sub Mediterranean bioclimates, moderate water stress, full flowering stage of the plant, and oven drying method can be critical factors that are able to enhance the percentage of carvacrol in the oil of the plants with chemotype I.

Chapter 5
Biological and Pharmacological Activity

5.1 Antibacterial Activity

Antibacterial property of *S. spicigera* oil against 25 plant pathogenic bacteria was tested and the results exhibited a broad spectrum of antibacterial activity attributed to the high content of carvacrol and thymol in the oil. Furthermore, it had bactericidal activity toward 14 strains of those tested bacteria. The hexane extract of the plant, which was rich in thymol and carvacrol, demonstrated lower antibacterial activity than its oil with respect to the lack of other minor terpenic constituents presented in the oil. Moreover, *S. spicigera* oil was more active against some seed borne pathogens than streptomycin sulfate that was used as the positive control [115]. The result of a study showed that among some Gram-positive and Gram-negative bacteria, the maximum inhibitory effect of the essential oil of *S. thymbra* were against *B. subtilis, S. maltophilia*, and *C. luteola* [116]. The antibacterial activity of *S. cuneifola* oil, as shown in Table 5.1, revealed the capacity of this oil for prevention of food born bacteria. This effect can be related to the presence of carvacrol, γ-terpinene and *p*-cymene [71]. The essential oil of *S. hortensis* has stronger and broader spectrum activity against tested bacteria in comparison with nonpolar fraction of methanol extract (Table 5.1). This can apparently be related to the high contents of carvacrol and thymol in the essential oil [86]. Antimicrobial activity of the methanol and hexane extracts of *S. hortensis* against 147 laboratory strains belong to 55 bacterial species, and 31 isolates of one yeast and four fungi species (including human and plants pathogens) were evaluated and the results indicated that the methanol extract was more potent than the hexane extract of the plant. In addition, clinical isolates of *Escherichia coli, Kocuria varians*, and *Micrococcus luteus* were found to be sensitive to the methanol extract of *S. hortensis* suggesting that this extract can be used for therapy of human infections [117]. Inhibitory activity of the oils obtained from *S. parnassica* ssp. *parnassica* during the different stages of the plant growth (flowering and vegetative stages) was tested toward clinical strain of *Helicobacter pylori* in culture media. The oil of flowering stage showed greater anti-*H. pylori* activity than the oil of vegetative stage with the half maximal inhibitory concentration (IC_{50}) of 250 and 500 µg/mL, respectively. The oil of flowering stage was reach in carva-

S. Saeidnia et al., *Satureja: Ethnomedicine, Phytochemical Diversity and Pharmacological Activities,* SpringerBriefs in Pharmacology and Toxicology,
DOI 10.1007/978-3-319-25026-7_5

Table 5.1 Active extracts of some *Satureja* species against different types of bacteria

Plant names	Bacterial species	References
S. hortensis	*A. baumanii, B. amyloliquefaciens, B. cereus, B. macerans, B. megaterium, B. subtilis, B cepacia, C. michiganense, E. cloacae, E. fecalis, E. coli, K. pneumonia, P. vulgaris, P. aeruginosa, P. fluorescens, Ps. syringae, S. enteritidis, S. aureus, S. epidermis, S. pneumonia, S. pyogenes, X. campestris*	[86]
S. hortensis[a]	*B. subtilis, E. fecalis, P. aeruginosa, S. enteritidis, S. pyogenes*	[86]
S. montana	Methicillin-resistant *S. aureus*	[127]
	E. coli, E. coli O157:H7, *L. monocytogenes, P. aeruginosa, S. enteritidis, S. aureus, S. typhimurium, L. monocytogenes, S. flexneri, Y. enterocolitica, B. subtilis, S. cerevisiae, Acinetobacter calcoacetica, Brevibacterium linens, Brocothrix thermosphacta, Clostridium sporogenes, Lactobacillus plantarum, Leuconostoc cremoris, Micrococcus luteus, Salmonella pullorum, Vibrio parahaemolyticus, Plesiomonas shigelloides, Clostridium perfringens*	[85, 118, 119, 128–132]
S. thymbra	*B. subtilis, M. luteus, S. mutans, S. aureus, S. epidermidis, E. coli, P. stutzeri, S. maltophilia, C. luteola*	[116]
S. cuneifolia	*B. subtilis, E. faecium, L. monocytogenes, S. aureus, E. coli, P. mirabilis, P. aeruginosa, S. typhimurium, A. hydrophila, B. amyloliquefaciens, B. brevis, B. cereus, B. laterosporus, C. xerosis, E. faecalis, E. faecium*, E. coli, *K. pneumonias, M. luteus, M. smegmatis, P. vulgaris, Y. enterocolitica*, methicillin-resistant *S. aureus, Pectobacterium carotovorum* pv. *carotovorum, P. corrugate, P. fluorescence, P. savastanoi* pv. *glycinea, P. savastanoi* pv. *phaseolicolta, P. savastanoi* pv. *atrovafaciens, P. viridiflava, Xantomonas campestris* pv. *pruni, Bifidobacterium adolscentis, B. dentium, B. infantis, B. longum, B. pseudocatenulatum, Clostridium* spp., *Lactococcus* subsp. *lactis, L. lactis* subsp. *cremoris, L. lactis* subsp. *Diacetilactis*	[50, 71, 76, 133]
S. parnassica ssp. *parnassica*	*S. aureus, B. cereus, E. coli, H. pylori*	[110]
S. parvifolia[b]	*E. coli, S. aureus, P. aeruginosa, Shigella* ssp., *Streptocuccus* ssp.	[134]
S. parvifolia[a]	*Plasmodium falciparum*	[125]
S. brownei[c]	*S. aureus, S. pyogenes*	[135]
S. khuzistanica[a]	*S. aureus, S. epidermidis, E. coli, P. aeruginosa, S. typhi*	[136]
S. khuzistanica[d]	*S. aureus, P. aeruginosa*	[137]
S. sahendica	*S. aureus*	[138]
S. boissieri, S. coerulea, S. pilosa, S. icarica	*E. coli, S. aureus, P. aeruginosa, Enterobacter aerogenes, Proteus vulgaris, S. typhimurium*	[77]

[a] Methanol extract
[b] Total extract of flavonoids
[c] Ethanol extract
[d] Essential oil preparations (Dentol®)

crol (20.40 %), while the oil of vegetative stage contained less amount of carvacrol (1.59 %) [110]. Antibacterial activity of the essential oil of *S. montana* against food born bacteria like *L. monocytogenes* and *E. coli* makes it suitable alternative instead of synthetic chemical preservatives in food commodity [118].

As a matter of fact, the essential oil of *S. montana* strongly inhibited the enteropathogens including *E. coli, Plesiomonas shigelloides, Shigella flexneri, Salmonella enterica* serov *typhimurium, Yersinia enterocolitica*, and *Vibrio parahaemolyticus*, which were isolated from patients with enteric infections. The results indicated that the above mentioned oil may be effective in the enteric infections and warrants further investigation [85]. Antibacterial activity of the *S. montana* oil with high content of thymol (28.99 %) was analyzed toward *Clostridium perfringens* type A inoculated in sausages with different levels of sodium nitrite (0–200 ppm) for 30 days. *In vitro* assays showed that the oil caused structural damage and cell lyses in the tested bacterium. Synergism effect was also observed between the essential oil and the synthetic additive [119]. The essential oils of different ecotypes of *S. khuzestanica*, possessing different amounts of carvacrol (42.5–94.8 %), were evaluated for their antibacterial activity against four pathogens namely *S. aurous, B. cereus, E. coli*, and *P. aeroginosa*. The results showed that the oil with highest content of carvacrol inhibited the bacteria more strongly [120]. The oil increased permeability of the cell membrane of bacteria, causing release of the cell constituents and decreasing the ATP concentration in the bacteria cells as well as intracellular pH [121]. Terpenes in the essential oils are able to be penetrated or disrupt the lipid structures, where in cell membrane causing loose of membrane integrity and dissipation of the proton motive force. Carvacrol makes membrane permeable to potassium ions and protons leading to acidifying the cytoplasm, and suppresses the synthesis of ATP [122–124]. Interestingly, the methanol extract of *S. parvifolia* presented high anti-plasmodial activity with IC_{50} value of 3 μg/mL comparable with *Artemisia annua* [125]. However, different extracts of *S. parvifolia* (methanol, dichloromethane and hexane extracts) were not effective against some bacteria and fungi *in vitro* [126].

5.2 Antifungal Activity

Previous studies mostly concentrated on inhibitory effects of various essential oils obtained from *Satureja* species against fungi. Here, we provided an overview on such studies, where the tested extracts of the plants and MIC values of those extracts have been summarized in Table 5.2. It is reported that the essential oil of *S. montana* with concentration of 1 % significantly inhibited the growth of both *Botrytis cinerea* and *Penicillium expansum* in post-harvest control of apples similar to the chemical control used, tebuconazole, after 15 days [94]. It is also reported that the essential oil of *S. thymbra* contained considerable amounts of phenolic compounds (thymol and carvacrol) and exhibited strong inhibitory effect against *Fusarium moniliforme, Rhizoctonia solani, Sclerotinia sclerotiorum, Phytophthora capsici*. The above mentioned chemicals have been considered as the fungi-toxic compounds of

Table 5.2 Active extracts of some *Satureja* species against different types of fungi and their MIC values

Plants name	Microorganisms	MIC[d]	References
S. hortensis	*C. albicans*	250	[86, 143]
	A. alternate	62.5	
	A. flavus	31.25	
	A. variecolor	125	
	F. culmorum	125	
	F. oxysporum	250	
	Penicillium spp.	125	
	Rhizopus spp.	250	
	R. solani	125	
	M. fructicola	31.25	
	T. rubrum	31.25	
	T. mentagrophytes	62.5	
	M. canis	62.5	
	S. sclerotiorum	125	
	S. mino	250	
S. hortensis[a]	*C. kefyr*	–	[144]
S. thymbra	*F. moniliforme, R. solani*	–	[75, 116, 139, 145]
	S. sclerotiorum	–	
	P. capsici, C. albicans	–	
	Mycogone perniciosa	–	
S. hortensis	*A. flavus*	6.25	[146, 147]
	A. parasiticus	–	
S. montana	*F. poae*	150	[148]
	F. equiseti	150	
	F. graminearum	100	
	F. sporotrichoides	100	
	F. culmorum	250	
	A. solani	300	
	R. solani	250	
	P. cryptogea	250	
	B. cinerea	250	
	S. parasitica	200	
S. cuneifolia	*A. fumigatus*	500	[50, 76, 133]
	C. albicans	60	
	C. sake	400	
	S. cerevisiae	120	
	Kluyveromyces marxianus	1000	
	Pichia membranaefaciens	400	
	Saccharomyces cerevisiae	400	
	Schizosaccharomyces japonicas	400	
	Schizosaccharomyces pombe	400	
	Torulospora delbruekii	400	
	Zygosaccharomyces bailii	400	

Table 5.2 (continued)

Plants name	Microorganisms	MIC[d]	References
S. khuzistanica[b]	*A. niger*	2000	[136]
	C. albicans	1000	
S. khuzistanica	*A. flavus*	1250	[149]
	A. niger	2500	
	Penicillium	625	
	Fusarium	625	
	Alternaria	625	
	Rhizopus	625	
	Mucor	625	
S. sahendica	*C. albicans*	3.125	[138]
S. mutica	*C. albicans*	1333	[142]
	S. cerevisiae	1333	

[a] spice
[b] methanol extract
[c] aq. ethanol (80 %) extract
[d] MIC (Minimum Inhibitory Concentration) values express as μg/mL

the oil [139]. It is believed that these phenolic compounds show antifungal activity on cell membranes causing leakage of intracellular metabolites [140].

Furthermore, it was found that although the oil of *S. hortensis* had antifungal activity higher than amphotericin B (used as the positive standard), its methanol extract did not show antifungal effect. Antifungal activity of the oil might be attributed to the high concentration of the phenolic compounds including carvacrol and thymol in the oil [86]. Moreover, hydrosols of *S. hortensis* (at the dose of 15 %) exhibited fungicidal activity with 100 % inhibition of mycelial growth toward some plant pathogens including *R. solani, B. cinerea* and *A. citri*. As a matter of fact, this spice plant attracts a particular interest to be applied in the food, storage products and cosmetic industries [141].

Additionally, the essential oil of *S. mutica* examined against some filamentous fungi including *Aspergillus niger, Trichophyton rubrum, Trichoderma reesei* and *Microsporum gypseum* using poisoned food technique. The MIC value of the oil against all the tested fungi were assessed as $\geq$0.25 μL/mL [142]. Spore germination of *Afternaria solani, Sclerotium cepivorum, Colletotrichum coccodes* were inhibited by 1 % of *S. parvifolia* oil, although the spores germinated when they were transferred to an oil free medium. That means the essential oil caused reversible inhibition, and did not lyse the spores. Furthermore, the mentioned volatile oil with concentration of 1 % caused complete inhibition on mycelial growth, while did not cause mycelial death, since the mycelia of the fungi were able to grow when transfered into the oil- free medium [100]. Furthermore, the essential oils of *S. boissieri, S. coerulea, S. pilosa* and *S. icarica* showed spore inhibition against *Penicillium canescens* after 3 days incubation, whereas they did not inhibited *A. niger, Penicillium steckii* and *P. sublateritium* germination [77]. Generally, the growth of some tested fungi were inhibited in presence of the essential oil of *Satureja* species, however the results of some other studies revealed that the essential oils had no fungicidal activity toward the fungi.

5.3 Antiviral Activity

Antiviral activity of *Satureja* species were examined against some viruses. Literature showed that the aqueous extract of *S. montana* possessed a potent anti-HIV-1 effect, and also this extract could inhibit the giant cell formation in co-culture of Molt-4 cells with or without HIV-1 infection. This means its inhibitory property against HIV-1 reverse transcriptase is specified [150]. The essential oil of *S. montana* ssp. *variegata* showed antiviral activity toward Tobacco Mosaic Virus (TMV) and Cucumber Mosaic Virus (CMV). When the oil was applied on the hos, the number of lesions reduced to 29.2 % for TMV infection and 24.1 % for CMV infection. Thymol and carvacrol were also examined on phytovirals, where inhibitory effect of thymol was stronger than carvacrol. Comparison of the percentage of inhibition suggested that there is no synergistic effect between thymol and carvacrol in antiviral activity of the oil [74]. Additionally, the extract of *S. boliviana* was active against both herpes simplex type I (HSV-1) and vesicular stomatitis virus (VSV) [151].

5.4 Anti-leishmania Activity

Surprisingly, the essential oil of *S. punctata* strongly inhibited promastigotes and axenic amastigote forms of *Leishmania donovani* and *L. aethiopica*, and exhibited high cytotoxicity and hemolytic activity. It is concluded that both active and inactive constituents of the oil play role in the mentioned effects. Active compounds showed synergistic effects, while inactive ones increased absorption and bioavailability of the active compounds [106].

5.5 Antitrypanosoma Activity

Different fractions of *S. macrantha* and *S. mutica* including acetone, methanol, and water fractions of the plants were tested against *Trypanosoma cruzi*, the ethiological agent of Chagas disease, of which acetone fractions of both plants were observed to be the most active extracts. Preliminary phytochemical investigation showed that the active fractions were rich of flavonoids and terpenoids [152].

5.6 Insecticidal Activity

The volatile oil of *S. hortensis* caused a high mortality against the nymphs and adults of *Tetranychus urticae* Koch and adults of *Bemisia tabaci*, which are worldwide economic pests in both the field and greenhouse. The results showed that mortality increased while exposure dose and time were enhanced [153]. Furthermore, larvae of the tobacco cutworm, *Spodoptera litura*, were topically administered together with the essential oil of *S. hortensis* and the results suggested that the oil was highly toxic to the cutworm. Thymol and carvacrol, major constituents of the oil, have most probably been accounted for insecticidal activity of the volatile oil [154]. In addition, the essential oil of *S. hortensis* showed a high mortality against mosquito larvae *Culex qiunquefasciatus* (LC_{50}: 36.1 μg/mL) with a short-term exposure in water contaminated by lethal doses of the oil. Mortality of the larvae increased significantly in relation to the exposure time. In addition, total mortality at the end of their development was about $75.1 \pm 6.9\,\%$. Total emergence of adult for control was 77.3 %, while it was 16.0 % for the oil of *S. hortensis* [155].

In another study, the essential oil of *S. thymbra* showed insecticidal activity against *Drosophila melanogaster* larvae with LD_{50} value as 3.3 μg/mL. The amount of each compound that allowed 15 % of the larvae to develop to the adult stage (LD_{50}) was estimated. The LD_{50} values of the plant essential oil, thymol and carvacrol were calculated as 3.3, 2.6 and 1.6 μL/mL, respectively. However, the results demonstrated that the insecticidal activity of the plant oil and its major constituent is not linearly dependent. Evaluating the toxicity of a mixture of thymol and carvacrol (two main constituents of the oil) suggested that there is an antagonistic phenomenon for these phenolic compounds. Therefore, the effect of the plant oil may be attributed to the effect of other compounds or possible synergism effect [156].

5.7 Antioxidant Activity

In the literature, different approaches have been mentioned for determination of antioxidant properties of herbal extracts resulted in dispersed findings, which are conflicting and hardly comparable. Following, some of the most important findings are summarized in Table 5.3. Among different studies, some investigations were carried out on antioxidant activity of *Satureja* species regarding the usage in food commodity. For instance, dried leaves of *S. hortensis* significantly showed antioxidant activity in dressing products more than propyl gallate that is a standard antioxidant in this type of product [157]. Free radical scavenging evaluation was performed on different extracts of *S. hortensis* along with methanol extract of the plant callus. The results showed that the strongest activity toward free radicals was about $IC_{50} = 23.76 \pm 0.80$ μg/mL, which is comparable to positive standard butyl hydroxyl toluene (BHT) with IC_{50} value of 19.80 ± 0.50 μg/mL. The order of activity for other extracts of the plant was as follows: aqueous fraction of methanol extract

Table 5.3 Antioxidant activity of some *Satureja* species using different methods

Plant name	Extract	Methods of evaluation	Inhibition (%)	IC_{50} (μg/mL)	Ref.
S. sahendica	Essential oil	Free radical scavenging against DPPH	–	7.85±0.06[a]	[138]
				8.34±0.06[b]	
				8.12±0.07[c]	
S. cilicica	Essential oil	Free radical scavenging against DPPH	–	32.02±0.58	[167]
		Phosphomolybdenum method		101.16±3.32	
S. Montana	Essential oil	Tyrosine nitration induced by peroxynitrite		43.9	[87, 165]
		β-carotene bleaching (inhibition of linoleic acid oxidation), thiobarbituric acid method		–	
		Malondialdehyde formation induced by peroxynitrite		27.2	
S. mutica	Aqueous methanol extract (80 %)	Free radical scavenging against DPPH	93.39±2.55	–	[166]
		Xanthine-oxidase activity	55.96±1.28		
		Inhibition of lipid peroxidation by the ferric thiocyanate	31.15±0.39		
		Pro-oxidant effect	56.8		
S. hortesis	Essential oil	Free radical scavenging against DPPH	75.2±0.6[d] 49.9±0.4[e]	–	[20]
	Ethanol extract		95.8		
	Acetone extract		33.0		
	Aqueous extract	Fenton reaction	35.4±3.4		[168]
S. cuneifolia	Essential oil	β-carotene bleaching (inhibition of linoleic acid oxidation)	84.5±0.3		[79]
	Methanol extract		95.2±0.2		
S. spicigera	Essential oil	β-carotene bleaching (inhibition of linoleic acid oxidation)	81.7±1.14		[72]
	Methanol extract		65.9±1.77		

[a] oil obtained from pre-flowering stage
[b] oil of flowering stage
[c] oil of post flowering stage
[d] Lithuanian origin
[e] Bulgarian origin

>chloroform fraction of methanol extract >essential oil. In contrast, the oil of *S. hortensis* inhibited linoleic acid oxidation (95 %) in compared to the chloroform extract (90 %). High activity of the essential oil seems to be related to the high content of thymol, carvacrol and γ-terpinene in the oil [86].

Sunflower oil contained 0.5 % ethanol extract of *S. hortensis* was reported to be stabilized effectively more than those contained 0.02 % of BHT. This result indicated that the above mentioned herbal extract is suitable antioxidant for stabilizing sunflower oil [158]. Furthermore, the ethyl acetate extract of *S. hortensis* was found as the most active fraction of the plant extract. Therefore it is suitable to retard free radical-mediated degradation of susceptible components [159]. The results of a study revealed that stable free radicals can be created from phenolics (carvacrol and thymol) in the oil of *S. hortensis* through reaction with O_2^- and hydrogen atom donation to form stable paramagnetic species, therefore these compounds can control lipid peroxidation in the membrane of the plants [160]. Moreover, the ethanol extract of *S. hortensis* improved oxidative and heat stability of sunflower oil in a dose dependent manner [161].

An extract of *S. montana* presented high antioxidant activity in hemodialysis assay *in vitro* and less than 10 % of hemodialysis occur during 4 h incubation with H_2O_2 Red blood cell model. In addition, the extract of *S. montana* showed antioxidant property, and also important protection against H_2O_2 upon the phage-mediated infection in the bacteriophage P22/*Salmonella Typhimurium* system [162]. Results of another study indicated that the volatile oils of *S. montana* (oil obtained by SFE and HD methods) strongly scavenged free radicals and inhibited lipid oxidation more than the extract that obtained using Soxhlet method [163].

Moreover, antioxidant activity of the essential oil of *S. montana* and *S. subspicata* were examined by DPPH test. The effectiveness was comparable with thymol, which was used as a positive control. The oil of *S. subspicata* was more active in reducing stable DPPH radical attributed to the high content of thymol and carvacrol [164]. Regarding the results presented in Table 5.3, *S. montana* oil inhibited formation of 3-nitrotyrosine and malondialdehyde that might be due to its high content of carvacrol [165]. The aqueous methanol extract (80 %) of *S. mutica* with concentration of 1 mg/mL inhibited free radicals 93.39 ± 2.55 % in comparison to BHT (96.47 ± 1.61 %) [166].

In another study, antioxidant activity of *S. montana* was analyzed in various extraction times and plant particle sizes. The results indicated that antioxidant power of the plant increased by increasing the extraction time and decreasing the particle size. This means that an increase in time and surface area of the plant material caused more mass transfer between the plant and solvent [22]. Free radical scavenging activity of the methanol extract and essential oil of *S. cuneifolia* were determined and IC_{50} values of those were calculated as 26.0 ± 1.2 and 65.1 ± 2.2 μg/mL, respectively. However, phenol contents of the methanol extract was evaluated more than those for the essential oil (222.5 ± 0.5 and 185.5 ± 0.5 μg/mL, respectively) [79].

Furthermore, free radical-scavenging capacities of the extracts and oils of *S. spicigera* and *S. cuneifolia* were measured in DPPH and β-carotene-linoleic acid assays. In comparison with the standard compounds including BHT, ascorbic acids, curcumin, and α-tocopherol, both oils and extracts considerably exerted the antioxidant activity [72]. Peroxynitrite, $ONOO^-$, is considered as a relevant radical concerning with pathological and toxicological process, since radicals ($NO_2^\bullet$ and $OH^\bullet$) formed from its degradation causing lipid peroxidation, disruption of cellular structures, inactivation of enzymes and ion channels through protein oxidation and nitration, and DNA damages. The essential oil of *S. cilicica* with concentration of 2 % could effectively reduce the oxidation of butter, and thus this oil can be a source of natural antioxidant and aroma for butter [167].

5.8 Allelopatic Property

Allelopatic activity of *S. montana* oil was examined on some weeds and crops to evaluate their potential as germination inhibitors. The oil with 57 % carvacrol completely inhibited both crops and weeds germination [169].

5.9 Cytotoxicity

Bioassay guided isolation of the active compounds of *S. gilliesii* afforded sesquiterpenes namely (+)-T-cadinol and (−)-cadin-4-en-1-ol, which showed high toxicity with LC_{50} values of 7.4 and 6.2 ppm, respectively. However, isolated monoterpenes, acetylsaturejol and isoacetylsaturejol exhibited toxicity at level of 100 ppm [36]. An ethanol extract of *S. montana* was used to evaluate its potential anti-tumor effect against Neuro-2a cells. The LC_{50} value of the extract was assessed as 2.56 mg/mL against examined cells indicating that the plant possesses relatively weak antitumor effect [170]. Some hepatoma cell lines were divided into two groups of HBV (+) and HBV (−), and treated with decoction of *S. hortensis*. The extract of the plant showed significant inhibitory activity on three HBV (−) cell lines (HepG2/C3A, HA22T/VGH and SK-HEP-1) in a dose-dependent manner. In the group of HBV (+) cell lines (including Hep3B and PLC/PRF/5), the cytotoxicity of the extract was lower than HBV (−) of hepatoma cell lines [171]. Cytotoxicity of some flavonoids from *S. atropatana* were tested on *Artemia salina* larva, in which the tested compounds showed toxicity less than the positive control, berberine hydrochloride [31].

5.10 Genotoxicity

The somatic mutation and recombination test on *Drosophila melanogaster* revealed that *S. thymbra* oil was not genotoxic, while thymol, the major components of the oil, showed genotoxic but not recombinagenic activity. The genotoxic activity was tested using the wing somatic and recombination tests (SMART) in *D. melanogaster*. A tested material may reduce or increase mutation rate depending on the genotoxicity potential or may be even be ineffective. These effects are expressed as mosaic spots on the wings of trans-heterozygous female flies. The concentration of each compound that allowed 50 % of the *Drosophila* larvae to develop to adult stage (LD_{50}) was determine [156].

5.11 Prevention of Oxidative Degradation of DNA and Deoxyribose

Different boiling and infusion extracts of *S. montana* were tested but anti- or pro-oxidant protection of DNA was not observed even by increasing the volume from 200 to 400 μL, while high degree of deoxyribose protection was observed with leaf boiling extract of the plant (72.05 %) [172].

5.12 Anti-Diabetic Activity

Methanol, hexane and dichloromethane extracts of *S. hortensis* increased insulin-stimulated glucose uptake dose dependently, while *S. montana* extracts had no effect on glucose uptake. To ensure that the results were not compromised by cytotoxic activity of the extracts, the insulin-stimulated glucose uptake test had been supplemented to the cytotoxicity analysis toward macrophages and endothelial cells. On the other hand, the extracts of both mentioned plants activated peroxisome proliferator-activated receptor (PPARs), a key regulator of adipogenesis and glucose homeostasis regarding their terpenoids, which makes them suitable sources for prevention and retardation of type 2 diabetes [173].

The results of another study showed that a dichloromethane extract of *S. montana* activated PPAR-γ dose-dependently *in vitro*. Despite of PPAR-γ activation *in vitro*, administration of the extract to mice did not show blood glucose lowering effect in compared to rosiglitazone [174]. Moreover, administration of *S. khuzestanica* had no effect on blood glucose level. It decreased phosphoenolpyruvate carboxykinase (PEPKC) and glycogen phosphorylase (GP) activity by 26 % and 24 % (of control), respectively. The authors concluded that the plant oil can stimulate glycogenolysis, and thus depletes hepatic glycogen storage compensating by blood glucose, since gluconeogenesis in liver is occluded. The mechanism for anti-diabetic activity of the oil may be attributed to its antioxidant activity [175]. Administration of *S. khuzestanica* caused lowering in fastening blood glucose and triglyceride levels in diabetic and hyperlipidemic rats [176].

5.13 Anti-Hyperlipidemia Activity

Pancreatic lipase is a key enzyme for absorption of dietary triglycerides, and therefore some herbal extracts were screened for their probable lipase inhibitory activity in the presence of *p*-nitrophenylpalmitate (PNP) and 5-bromo-4-chloro-3-indoxylpalmitate (X-pal) as substrates. The results implied that the extract of *S. montana* inhibited the lipase in presence of PNP and X-pal ranking less than 40 % and 40–70 %, respectively, while, the extract of *S. hortensis* inhibited the enzyme below 40 % in the presence of both substrates [177]. Administration of flavonoid fractions of *S. hortensis* with dose of 10 mg/kg to rabbits for 8 weeks resulted in a significant prevention of diet-induced rise of serum cholesterol [178]. Furthermore, patients with type 2 diabetes (12 males and 9 females) were randomly divided into two groups, in which 11 patients were received *S. khuzestanica* tablet (250 mg flowering aerial parts per tablet) and 10 patients just received placebo. Total cholesterol and low-density lipoprotein-cholesterol (LDL-C) levels decreased, while high-density lipoprotein-cholesterol (HDL-C) and total antioxidant power increased after 2 months in comparison with placebo group, which showed no changes. However, other parameters like triglyceride, creatinine, blood glucose and thiobarbituric acid reactive substance (TBARS) did not significantly change in the treatment group. Although it is concluded that the plant could be used as supplements in diabetic patients with hyperlipidemia, more observations are required in a larger group of patients with longer time of treatment [179].

5.14 Inhibition of Angiotensin Converting Enzyme (ACE) and Digestive Enzymes

The chloroform extract of *S. thymbra* weakly inhibited α-amylase and α-glucosidase activities with IC_{50} values of 351.6 and 289.8 μg/mL, respectively, whereas the inhibitory activity against ACE was reported by IC_{50} value more than 150 μg/mL [180].

5.15 Anticholinesterase Activity

Different types of *S. montana* extract obtained by hydrodistillation, soxhlet and supercritical fluid extraction methods were evaluated for its possible activity on acetyl- and butyryl-cholinesterase, two important enzymes in control of Alzheimer's disease. Supercritical extract of *S. montana* was rich of (+)-catechin, chlorogenic, vanillic and protocatechuic acids, and significantly inhibited butyryl-cholinesterase. On the other hand, nonvolatile conventional extract of the plant did not show activity toward the enzymes [181].

5.16 Vasodilation Activity

Aqueous extracts of two varieties of *S. obovata* Lag. subsp. *obovata*: var. *valentina* and var. *obovata* inhibited the contraction induced by acetylcholine and $CaCl_2$ in rat duodenum, as well as those induced by noradrenaline and $CaCl_2$ in rat aorta dose dependently. Both extracts exhibited relaxant effects in tissue pre-contracted with K^+ and aorta of rat pre-contracted with noradrenaline. Vasodilatory activity of both extracts decreased by removing endothelium, however the variety *"valentine"* exerted stronger inhibitory activity in all tested tissues. The mechanism of action may be complex due to the presence of different compounds in the extracts [182]. Bioassay-guided isolation of vasodilator fraction of *S. obovata* resulted in purification of three flavonoids named naringenin, eriodictyol, and luteolin, which all relaxed contractions induced by noradrenaline and KCl in isolated rat aorta [183]. The results of another study suggested that luteolin and eriodictyol inhibited both protein kinase C (PKC) and calcium influx involved in tonic-I and tonic-II phases inhibition, respectively, while naringenin only suppressed PKC related to the tonic-I phase [184]. Eriodictyol (5, 7, 3′,4′-tetrahydroxyflavanone) reversed vasoconstriction effect of noradrenaline and KCl in thoracic aorta rings in a concentration-dependent way. The compound also suppressed $CaCl_2$ and phorbol-12-myristate-13-acetate induced contractions, however did not show any activity toward PKC. It seems that this effect occurred due to the inhibition of calcium influx or other enzymes subsequent to the PKC activation related to the activation of contractile proteins like myosin light chain kinase [185].

5.17 Anti-Nociceptive and Anti-Inflammatory

The essential oil of *S. thymbra* is rich of γ-terpinene, carvacrol, thymol and *p*-cymene, and showed significant dose-dependent inhibition of licking and biting of hind paw in formalin paw test in both early and late phases. However, the essential oil did not show anti-nociceptive effect in rat tail-flick test. In hot-plate test, only two doses of 100 and 200 mg/kg of the oil showed significant anti-nociceptive effect in rats. Additionally, administration of the oil did not inhibit paw swelling in carrageenan model. Formalin test is a reliable model for testing different classes of analgesic compounds. [103]. It is believed that this test comprise of two phases. In the early phase, direct chemical activation of nociceptive afferent fibers occurred, and in the late phase peripheral inflammatory is responsible. Centrally acting analgesics like morphine exhibits their analgesic activity in both phase [186, 187]. Therefore, inhibition of licking and biting of hind paw in formalin paw test in both early and late phases in formalin paw test suggested that the plant oil may have central action. This hypothesis is supported by the effect of the oil in the hot-plate test as well. Since this test is sensitive to central analgesics, there is a contradiction in the results of tail-flick test using the plant oil and this test is also predominantly involve in central mechanism [103]. The same study was done to evaluate potential anti-nociceptive and anti-inflammatory activity of *S. hortensis* oil and extracts. Although, the essential oil of *S. hortensis* just the same as *S. thymbra* oil did not change the tail flick reaction latency, hydroalcoholic extract, polyphenolic fraction and essential oil of *S. hortensis* have good anti-nociceptive effect in formalin test. Moreover, polyphenolic fraction and essential oil of the plant showed potent anti-inflammatory activity in carrageenan model [188]. An aqueous extract of *S. boliviana* exerted anti-inflammatory and cytoprotective activity in the previous study [189]. Analgesic activity of the essential oil of *S. cuneifolia* with high content of carvacrol (63 %) was evaluated using tail-flick test, which showed slight analgesic activity in comparison with positive control [190]. Administration of *S. khuzestanica* oil to mouse model of inflammatory bowel disease (IBD) caused reduction in the enzyme myeloperoxidase (MPO) and TBARS concentrations. The enzyme of MPO catalyzes oxidation and is found in neutrophils, monocytes, and macrophages. In IBD, the levels of neutrophils and consequently MPO enzyme increased in inflamed tissues. The plant oil with dose of 1500 ppm exhibited potential protection comparable to prednisolone that was revealed by biochemical, macroscopic and microscopic evaluations. The probable mechanism for this effect may be related to antioxidant, antimicrobial, anti-inflammatory, and antispasmodic potential of the plant oil [191].

5.18 Antispasmodic and Anti-Diarrheal Activity

The essential oil of *S. hortensis* suppressed ileum contraction induced by 80 mM of KCl in a dose dependent manner, while 72 μg/mL of the oil completely abolished a response to KCl. This effect was last as long as the tissue was presented in the bath and 30–60 min after washing. Various doses of 9–36 μg/mL of the oil inhibited acetyl choline (ACh) contraction responses, while a complete suppress of ileum contraction was achieved at 36 μg/mL. Furthermore, none of the pre-treated rats by the plant oil had no wet defecation after castor oil administration. Generally, the results indicated that the essential oil of *S. hortensis* is ileum relaxant and it can inhibit castor oil induced diarrhea [9].

5.19 Rhinosinusitis Treatment and Nitric Oxide Synthesis (NOS) Inhibition

Effect of an aqueous extract obtained from *S. hortensis* (250 mg/kg) was examined in rhinosinusitis model inducted by *S. aureus* in rabbits. The result of the study showed that the sinus mucosa thickened in saline control group and sub-epithelial edema with dilated capillaries were observed followed by epithelial desquamation and sever sub-epithelial infiltration of polymorphonuclears (PMNs). However, in the group treated by *S. hortensis*, only mild amount of PMNs infiltered and minimally-thickened-sub-epithelial space was observed. The level of NOS in the tested group was significantly lower than the control group (5.6 ± 1.8 and 9.1 ± 1.2 mIU/mg protein, respectively). Except NO_2^-, the concentration of $NO^\bullet$ and NO_3^- were also significantly lower in mucosal specimen of the treated group in comparison with the control group. This study declared the beneficial anti-inflammatory and NOS inhibition activity of the plant extract in rhinosinusitis [192].

5.20 Influence on Fertility

An essential oil obtained from *S. Khuzestanica* was administered to male rats for 45 days. The results showed an increasing in both potency and fecundity at two doses of 150 and 225 mg/kg in comparison with control group. It also improved fertility index and litter size along with a reduction of post implantation loss in mated females. The results showed that the serum level of LH and estradiol did not change in treated group, while FSH and testosterone significantly increased with the oil administration. The oil also augmented spermatogonium, spermatid cords, leydig cells, spermatozoids, and sertoli cells especially at two doses of 150 and 225 mg/kg. In addition, the weights of the testis, epididymis, and seminal vesicles increased due to the augmentation of leydig cell and germ cells, as well as higher rate of spermatogenesis. These effects may be attributed to the antioxidant activity of the plant oil [193]. Cyclophosphamide is an anticancer agent with toxicity in male reproduction system. Co-administration of *S. khuzestanica* oil with the cyclophosphamide reduced lipid peroxidation in plasma and testis. The oil enhanced total antioxidant power of plasma and testis when it was administered alone or in combination with the drug. Treatment with cyclophosphamide resulted in reduction of the weight of testes, epididymis, seminal vesicle, and ventral prostate as well as decrease in sperm count and motility, whereas all changes were restored by the plant oil co-treatment followed by inhibition of DNA damage induced by cyclophosphamide. Fecundity and fertility indices and litter size in treated animals were improved by administration of the oil. The beneficial effect of the oil could be attributed to its antioxidant activity [194].

5.21 Inhibition of Hemorrhagic Cystitis

Literature revealed that microscopic hemorrhage occurred in urinary bladder with administration of cyclophosphamide in rats, whereas co-administration of *S. khuzestanica* oil and cyclophosphamide resulted in normal urinary bladder. The plant oil also recovered plasma lipid peroxidation and total antioxidant power, which was altered by cyclophosphamide. Prevention of mast cells accumulation and improvement of oxidative stress are proposed as the mechanism of the plant oil beneficial effect [195].

5.22 Cytoprotective Activity

Cytoprotective activity of different extracts of *S. boliviana* has been demonstrated in ethanol-induced ulcer in rats. An aqueous extract of the plant that contained flavonoids, tannins and saponins showed the highest activity (52.8 %). However, ethanolic extract with similar composition of secondary metabolites showed the weakest cytoprotection activity (5.7 %) among the tested extracts. The hexane and dichloromethane extracts exerted cytoprotective effect at levels of 25.7 and 14.3 %, respectively [189]. It is reported that co-administration of *S. kuzestanica* oil with malathion, an organophosphorus (OP) pesticide, reconstructed induction of mitochondrial glycogen phosphorylase (GP) and phosphoenolpyruvate carboxykinase (PEPCK) in hepatic cells. Malathion causes glucose release from hepatic cells into blood trough stimulation of glycogenolysis and gluconeogenesis, which interferes with *S. khuzestanica* oil by its antioxidant potential and increases acetylcholine esterase activity in rats [196].

An ethanolic extract of *S. hortensis* almost completely eliminated radicals, which were produced in presence of H_2O_2 in Jurkat cell culture. In addition, ethanolic extract of the plant significantly reduced oxidative stress in the cell culture. The aqueous extract and rosmarinic acid-containing fraction exhibited antioxidant activity but in remarkable lower amount in comparison to the ethanolic extract. Although all the mentioned extracts increased viability of Jurkat cell in the presence of hydrogen peroxide, aqueous extract of the plant was most effective for restoring cell viability. These effects may be attributed to the radical scavenging activity of the phenolic compounds or activation of antioxidant enzymes, and release of anti-inflammatory agent like IL-10 [24]. Furthermore, essential oil and ethanolic extract of *S. hortensis* reversed DNA damage caused by H_2O_2 (as an inducer of oxidative stress) in rat lymphocytes. The observed antigenotoxic effect may be related to the presence of antioxidant compound in the oil and the extract [197].

Chapter 6
Satureja Bachtiarica: Phytochemistry and Pharmacology

6.1 History and Bibliography

Satureja belongs to the family Lamiaceae (subfamily: Nepetoideae) including annual aromatic plants, of which *S. bachtiarica* is an endemic Iranian species of *Satureja*. This species is widely distributed in the southern region of Iran [198, 199]. A literature review reveals several pharmacological effects reported from various species of this genus such as antiviral [151], antiprotozoal [200, 201], anti-diarrheal and anti-spasmodic [9], antibacterial, antifungal [202] and cytotoxic [31, 203] activities. The *in vitro* leishmanicidal effects of ethanolic and methanolic extracts of *S. khuzestanica* leaves on *Leishmania major* were previously evaluated and resulted in considerable growth inhibition of the parasite against *L. major* promastigotes (IC_{100} = 2.4 and 4.8 mg ml^{-1} and IC_{50} = 0.3 and 0.6 mg ml^{-1}, respectively) compared to glucantime as positive control, which inhibited the growth of *L. major* promastigotes with IC_{50} = 10.6 mg ml^{-1} [204].

Previous phytochemical investigations revealed the presence of thymol and carvacrol in the essential oils of many *Satureja* species [88, 205]. Ursan and oleanan triterpenoids, flavonoids, chalcones and sterols have been previously reported from the ethyl acetate and methanolic extracts of *S. macrantha, S. atropatana, S. spicigera* and *S. sahendica,* which grow wildly in Iran [31, 203, 206, 207]. There is only one paper about the antibacterial and chemical constituents of the essential oil of *S. bachtiarica*, indicates the presence of phenol (37.36 %), thymol (22.65 %) and *p*-cymen (19.29 %) as the major compounds. The antibacterial property of the volatile oil of *S. bachtiarica* may be mostly attributed to the phenolic compounds of the essential oil [208]. In this study, we aimed to report the isolation and identification of the main compounds of the ethyl acetate and methanolic extracts of *S. bachtiarica*, which has not previously been reported.

S. Saeidnia et al., *Satureja: Ethnomedicine, Phytochemical Diversity and Pharmacological Activities,* SpringerBriefs in Pharmacology and Toxicology,
DOI 10.1007/978-3-319-25026-7_6

6.2 Plant Material and Experimental Procedure

Aerial parts of *S. bachtiarica* Bunge were gathered from Chaharmahal-o Bakhtiari province (west of Iran) at its full flowering stage in September 2009. A voucher specimen of the plant was deposited at the Herbarium of the Institute of Medicinal Plants, ACECR, Tehran, and the plant specimen was identified by Mr. Yousef Ajani from the mentioned institute.

The ^{1}H and ^{13}C-NMR spectra were measured on a Brucker Avance TM 500 DRX (500 MHz for ^{1}H and 125 MHz for ^{13}C) spectrometer with tetramethylsilane as an internal standard and chemical shifts are given in δ (*ppm*). The MS data were recorded on an Agilent Technology (HP TM) instrument with 5973 Network Mass Selective Detector (MS model). The silica gel $60F_{254}$ pre-coated plates (Merck TM) were used for TLC. The spots were detected by spraying anisaldehyde-H_2SO_4 reagent followed by heating (120 °C for 5 min).

6.3 Isolation Process

The dried and flowered aerial parts of *S. bachtiarica* (2 kg) was cut into small pieces and extracted three times with ethyl acetate and methanol, consequently, at room temperature to obtain ethyl acetate (38 g) and methanol extracts (50 g). The ethyl acetate extract was submitted to silica gel column chromatography (CC) with hexane: $CHCl_3$ (7:3), $CHCl_3$: AcOEt (7:3) and AcOEt as eluent to give six fractions (A-F). The fraction B (4.9 g) was subjected to silica gel CC with n-hexane: $CHCl_3$ (8:2) as an eluent to obtain three fractions (B_1-B_3). Compound **51** (53 mg) was resulted from fraction B_2 after silica gel CC chromatography with n-hexane: $CHCl_3$ (8:2). The fraction C (1.7 g) was submitted to silica gel CC with n-hexane: AcOEt (9:1, 7:3) to gain four fractions (C_1–C_4). Chromatography of the fraction C2 (167 mg) on a sephadex LH20 column, two times, with AcOEt: MeOH (4:6), resulted in isolation of the compounds **52** (15 mg) and **53** (13 mg).

The fraction E (11.2 g) was subjected to silica gel CC with n-hexane: AcOEt (8:2, 5:5) to afford five fractions (E_1–E_5). The fraction E_3 was submitted to silica gel CC with $CHCl_3$: AcOEt (9:1, 4:6), and then was chromatographed on sephadex LH_{20} with AcOEt: MeOH (3:7) to yield compound **54** (12 mg).

The MeOH extract was successively submitted to silica gel CC with AcOEt, AcOEt: MeOH (8:2, 1:1, 1:9) as eluents to result in five fractions M_1–M_5. The fraction M_2 was fractionated on sephadex LH_{20} with MeOH, three times, to obtain compound **55** (6 mg). The fraction M_3 was subjected to silica gel CC with $CHCl_3$: MeOH (7:3), and then submitted to sephadex LH_{20} CC with MeOH to afford compound **56** (18 mg).

6.4 Phytochemical Constituents Found in *S. bachtiarica*

From the aerial parts of *S. bachtiarica*, two flavonoids, one phytosterol, two monoterpenes and one phenolic acid were isolated and identified as thymol [201, 204], β-sitosterol [209, 210], *P*-cymene-2,3-diol [211], naringenin [212], luteolin [213] and rosmarinic acid [214, 215] based on spectroscopic spectra (^{1}H-NMR, ^{13}C-NMR, HSQC and HMBC) compared to the known standard compounds which reported in the literature (Fig. 6.1). HMBC correlations of the compound **3** have been indicated in Fig. 6.2. To the best of our knowledge, this is the first report on the isolation and structural elucidation of these compounds from the species *S. bachtiarica*.

Naringenin (54): white needle crystal. ^{1}H NMR (500 MHz, CD_3OD):δ (ppm), 7.31 (2H, *d, J*=8.4 Hz, H-2′ and H-6′), 6.82 (2H, *d, J*=8.4 Hz, H-3′ and H-5′), 5.90

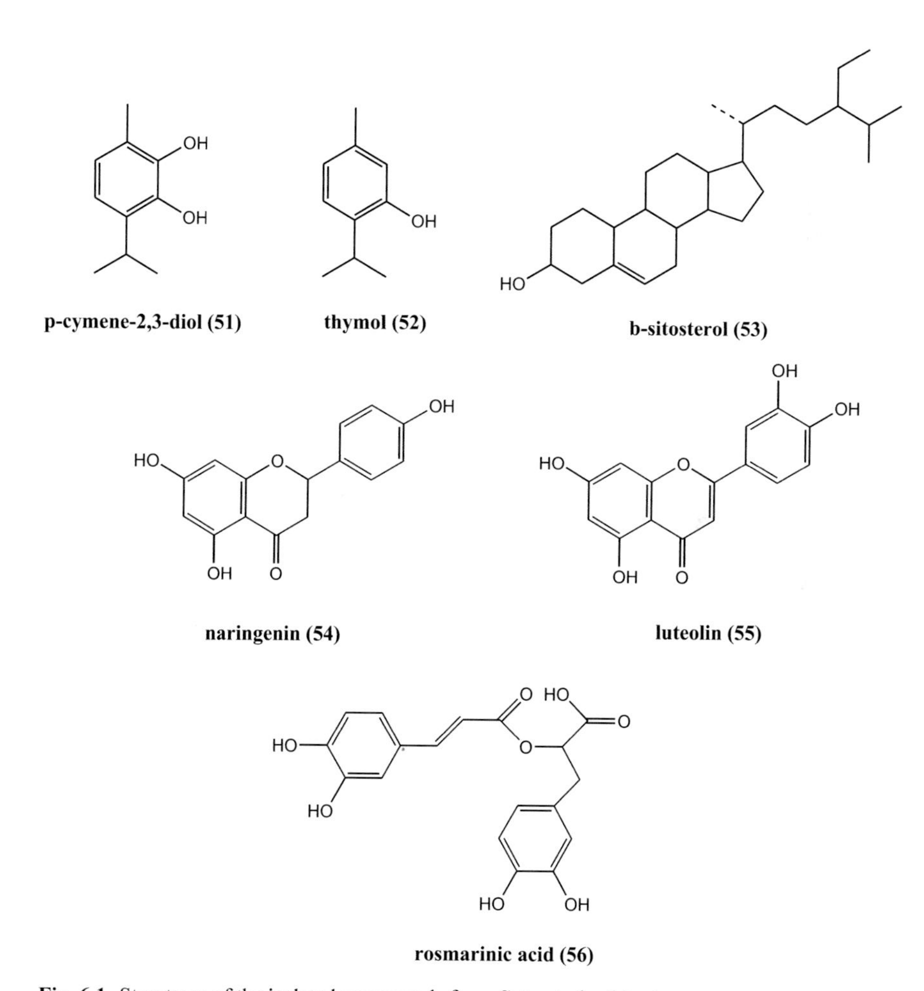

Fig. 6.1 Structures of the isolated compounds from *Satureja bachtiarica*

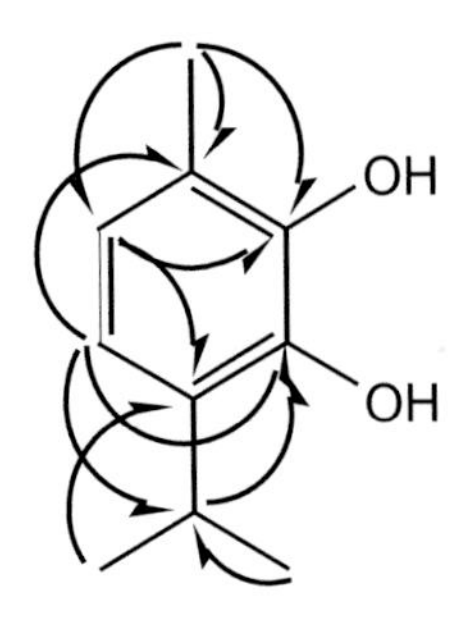

Fig. 6.2 The HMBC correlations in *P*-cymene-2,3-diol (H⇀C)

(1H, *brs*, H-6), 5.89 (1H, *brs*, H-8), 5.33 (1H, *dd, J*=12.9, 2.6 Hz, H-2), 3.11 (1H, *dd, J*=17.1, 12.9 Hz, H-3a) and 2.70 (1H, *dd, J*=17.1, 2.8 Hz, H-3b). ^{13}C NMR (125 MHz, CD_3OD):δ (ppm) 80.5 (C-2), 44.0 (C-3), 197.8 (C-4), 165.5 (C-5), 97.1 (C-6), 168.4 (C-7), 96.2 (C-8), 164.9 (C-9), 103.3 (C-10), 131.1 (C-1′), 129.0 (C-2′), 116.3 (C-3′), 159.0 (C-4′), 116.3 (C-5′), 129.0 (C-6′).

The ^{1}H-NMR and ^{13}C-NMR data of the p-cymene-2,3-diol and rosmarinic acid have been shown in Tables 6.1 and 6.2 respectively.

In continuation of our previous investigations on *Satureja*, the results of this study revealed the presence of thymol and *p*-cymene derivatives in *Satureja* genus. Naringenin, previously reported from *S. obovata* [216], is a weak phytoestrogen which exhibited partial anti-estrogenic activity in the female rat uterus and MCF-7 human breast cancer cells [217]. Rosmarinic acid, a bioactive phenolic compound, is found in many genus of Lamiaceae. Of which *Salvia, Melissa, Origanum, Lavandula, Rosmarinus, Thymus, Mentha, Perovskia, Zhumeria* and *Satureja* are the most important genus [21]. Among *Satureja hortensis, S. khuzestanica, S. bachtiarica, S. atropatana, S. mutica* and *S. macrantha*, the highest amount of rosmarininc acid has been reported in *S. mutica* (19.0 mg/g) and *S. hortensis* (16.3 mg/g), while it is reported as 5.7 mg/g of *S. bachtiarica* [21].

Luteolin, as one of the constituents of *S. parvifolia* had been previously reported as a cytotoxic flavone to various cancerous cell lines as well as the active components against *Plasmodium falciparum* K1 (IC_{50}=6.4 μg/ml) [218]. It is reported that

Table 6.1 NMR data of p-cymene-2,3-diol (δ values: ppm, $CDCl_3$)

No	δ_C	δ_H	HMBC
1	121.1	–	H-5, H-6, H-7
2	141.4	–	H-6, H-7
3	140.9	–	H-5, H-8
4	132.4	–	H-5, H-6, H-8, H-9 or H-10
5	117.4	6.67 (*d, J*=8 Hz, 1H)	H-8
6	121.9	6.70 (*d, J*=8 Hz, 1H)	H-7
7	15.3	2.23 (*s*,3H)	H-6
8	27.2	3.15 (*m*, 1H)	H-5, H-9 or H-10
9	22.6	1.25 (*d, J*=6.9 Hz, 3H)	H-8, H-10
10	22.6	1.25 (*d, J*=6.9 Hz, 3H)	H-8, H-9

Table 6.2 NMR data of rosmarinic acid (δ values: ppm, CD_3OD)

No	δ_C	δ_H	HMBC
1	128.0	–	H-5, H-7, H-8
2	115.7	7.03 (*brs*, 1H)	H-6, H-7
3	146.7	–	H-5
4	149.3	–	H-2, H-5, H-6
5	116.5	6.77 (*d*, *J*=8.0 Hz, 1H)	–
6	122.9	6.91 (*brd*, *J*=7.9 Hz, 1H)	H-2, H-7
7	146.6	7.50 (*d*, *J*=15.9 Hz, 1H)	H-2, H-6
8	115.1	6.27 (*d*, *J*=15.9 Hz, 1H)	–
9	169.1	–	H-7, H-8, H-8′
1′	131.2	–	H-5′, H-7′a, H-7′b, H-8′
2′	117.5	6.77 (*brs*, 1H)	H-6′
3′	145.9	–	H-5′
4′	144.7	–	H-2′, H-6′
5′	116.2	6.67 (*d*, *J*=8.0 Hz, 1H)	–
6′	121.7	6.63 (*brd*, *J*=8.1 Hz, 1H)	H-2′, H-7′a, H-7′b
7′	38.8	2.93 (*dd*, *J*=14.0, 9.8 Hz, 1H) 3.10 (*dd*, *J*=14.1, 2.5 Hz,1H)	H-2′, H-6′, H-8′
8′	77.8	5.09 (*dd*, *J*=9.7, 2.5 Hz,1H)	H-7′a
9′	177.6	–	H-8′

luteolin, as one of the constituents of *S. obovata* inhibited both tonic-I and tonic-II phases associated to the inhibition of Protein Kinase C (PKC) and calcium influx [183].

Taking together, there are a broad spectrum of the secondary metabolites in this plant that most of them are usual compounds in Lamiaceae family and particularly *Satureja* species. However, the importance of this endemic *Satureja* species should be considered regarding its effectiveness in Alzheimer's disease. Actually, *S. bachtiarica* exhibited a high protective effects against β-amyloid induced toxicity ($P<0.001$). The observed protective effects of *Satureja bachtiarica* were dose-dependent [219]. As a matter of fact, the major constituents of this plant mainly belong to polyphenolic and flavonoid compounds such as rosmarinic acid, naringenin and luteolin, which possessed antioxidant properties and may play a role in neuroprotection. Regarding the neuroprotective effect of this plant against β-amyloid induced toxicity, we recommend greater attention to their use in the treatment of Alzheimer disease.

Chapter 7
Discussion and Conclusion

The genus of *Satureja* consists of aromatic plants with traditional usage in treatment of nausea, indigestion, diarrhea, blood pressure, appetizing, cough, vomiting, toothache and externally for relieving rheumatoid pain and inflammation, scabies and itching along with emmenagogue and diuretic effects of the savory flowers [10, 11, 13]. Recent studies focused on isolation and purification of diverse secondary metabolites of this genus resulting mainly in identification of phenolic acid like rosmarinic acid, caffeic acid and chlorogenic acid [16, 20] followed by flavonoids [27], triterpenoids (such as OA and UA) [34, 35], sesquiterpenes [36] and iridoids [37]. A number of studies evaluated essential oil composition of *Satureja* species indicating monoterpens with *p*-menthane structure like carvacrol, thymol and *p*-cymene.

Different biological activities of the essential oils of these plants may be attributed to the high amount of the mentioned monoterpens such as antibacterial, antifungal, antiprotozoal, anti-leishmania, insecticidal and antioxidant activity. It is believed that the phenolic compounds may act on cell membranes causing leakage of intracellular metabolites [140]. On the other side, stable free radicals may create from phenolics (carvacrol and thymol) through reaction with O_2^- and hydrogen atom donation to form stable paramagnetic species, therefore these compounds can control lipid peroxidation in the cell membrane of plants' tissues [160]. Cytotoxicity and genotoxicity of these plants have been studied briefly indicating that some of the active secondary metabolites possess cytotoxicity, of them sesquiterpenes and thymol are the most considerable constituents [36, 156].

Evaluation of anti-diabetic activity of these plants shows controversial results, for instance: the extracts of *S. hortensis* activated insulin-stimulated glucose uptake dose dependently along with activation of PPARs, while *S. montana* activated PPARs but did not influence glucose uptake [173, 174]. Moreover, administration of *S. khuzestanica* have not influenced blood glucose level but decreased phosphoenolpyruvate carboxykinase (PEPKC) and glycogen phosphorylase (GP) activity [175].

Total cholesterol and low-density lipoprotein-cholesterol (LDL-C) levels decreased in diabetic patient with hyperlipidemia after two months administration of *S. khuzestanica*, while high-density lipoprotein-cholesterol (HDL-C) and total an-

S. Saeidnia et al., *Satureja: Ethnomedicine, Phytochemical Diversity and Pharmacological Activities,* SpringerBriefs in Pharmacology and Toxicology,
DOI 10.1007/978-3-319-25026-7_7

tioxidant power increased. Similarly flavonoid-containing fractions of *S. hortensis* prevented diet-induced rise of serum cholesterol [178, 179]. Positive influence of the plants on diabetic patient may be due to their beneficial effect on serum lipid profile relating to their inhibitory activity toward lipase enzyme or their antioxidant activity.

Vasodilation activity of the plants was successfully determined, and it seems that this effect occurs regarding to the inhibition of calcium influx or other enzymes subsequent to the PKC activation related to the activation of contractile proteins like myosin light chain kinase [184]. Anti-nociceptive effect of this plants are debatable, since they have positive effect in some tests like hot-plate and formalin paw tests, while did not show expected effects in rat tail-flick [103, 188, 190]. However, the mechanism of their anti-nociceptive effects can be ascribed to central action. Recent findings strengthen some traditional use of the plants of this genus for gastrointestinal disorders like antispasmodic and antidiarrheal activity due to inhibition of ileum contraction, antibacterial and anti-inflammatory activity of the plants. The probable mechanism for anti-inflammatory activity of *S. khuzestanica* in IBD model may be due to the antioxidant, antimicrobial, anti-inflammatory, and antispasmodic potential of the plant oil [191]. Favorable effects of the plants on fertility in the presence of harmful agent like cyclophosphamides or reduction of adverse effect of the drug like hemorrhagic cystitis may also be related to the antioxidant power of the plants extracts [193–195]. Presence of antioxidant compounds in the oils and extracts of the plants of this genus along with activation of antioxidant enzymes and release of anti-inflammatory agents like IL-10 are possible mechanisms for their cytoprotection activity [31, 89, 196, 197]. Taking together, the plants of this genus not only have long history of usage and excellent reputation in traditional medicine, but also could be a candidate source for finding new drugs in treatment of human disorders.

Appendix

Appendix (Tab. A.1)

S. Saeidnia et al., *Satureja: Ethnomedicine, Phytochemical Diversity and Pharmacological Activities,* SpringerBriefs in Pharmacology and Toxicology,
DOI 10.1007/978-3-319-25026-7

Table A.1 Chemical composition of essential oils of some *Satureja* spps

Plants name	*S. thymbra*									*S. spicigera*	*S. coerulea*		
Compounds name	Greece [152]	Turkey [99]	Turkey, Antalya (%v/w) [47]							Turkey [111]	Turkey [110]		
			Feb	March	April	May	June	July			July	Aug	Aug
α-thujene	–	1.98	–	–	–	–	–	–		–	–	–	–
α-pinene	3.80	1.47	4.2	4.3	3.2	2.4	4.3	4.1		3.01		0.05	0.3
Camphene	1.21	0.63	tr	0.5	tr	tr	tr	tr		–	–	–	–
Sabinene	0.10	–	–	–	–	–	–	–		–	–	–	–
β-pinene	1.20	0.47	2.2	2.5	1.0	tr	1.2	tr		0.34	–	–	–
1-octen-3-ol	0.03	–	–	–	–	–	–	–		–	0.3	0.1	1.2
Myrcene	1.50	1.52	5.1	5.1	7.2	5.6	8.7	8.9		1.92	1.3	–	4.4
3-octanol	0.10	–	–	–	–	–	–	–		–	–	–	–
α-phellandrene	0.07	0.35	–	–	–	–	–	–		–	–	–	–
α-terpinene	1.55	–	tr	tr	tr	tr	tr	tr		2.81	–	–	–
p-cymene	26.76	12.73	95.4	77.4	31.8	30.4	58.5	74.1		18.25	0.6	0.2	3.2
Limonene	0.70	0.78	tr	tr	tr	tr	tr	tr		–	1.8	0.3	5.1
β-phellandrene	0.05	–	–	–	–	–	–	–		–	–	–	tr
δ-3-carene	0.02	–	–	–	–	–	–	–		0.34	–	–	–
1,8-cineole	0.29	–	14.7	20.4	9.9	11.3	9.3	5.0		–	0.3	0.1	1.5
Cis-β-ocimene	0.02	–	–	–	–	–	–	–		–	0.9	tr	1.4
β-ocimene	–	0.58	–	–	–	–	–	–		–	–	–	–
Trans-β-ocimene	–	–	–	–	–	–	–	–		0.49	0.8	0.05	0.6
γ-terpinene	13.86	40.99	21.5	53.8	168.4	103.7	128.5	139.2		11.16	–	–	–
Cis-sabinene hydrate	0.08	0.40	–	–	–	–	–	–		–	0.4	0.5	0.7
α-terpinolene	0.09	–	–	–	–	–	–	–		–	0.06	–	–
Linalool	0.57	0.71	6.2	4.1	2.9	3.2	10.1	11.0		0.08	–	–	–
Trans-sabinene hydrate	0.49	–	–	–	–	–	–	–		–	1.0	1.6	2.2

Table A.1 (continued)

Plants name	*S. thymbra*								*S. spicigera*	*S. coerulea*		
Compounds name	Greece [152]	Turkey [99]	Turkey, Antalya (%v/w) [47]						Turkey [111]	Turkey [110]		
			Feb	March	April	May	June	July		July	Aug	Aug
Nonanal	–	–	–	–	–	–	–	–	–	–	0.04	–
Cis-*p*-menth-2-en-1-ol	–	–	–	–	–	–	–	–	–	0.08	0.1	0.2
Trans-pinocarveol	–	–	–	–	–	–	–	–	0.09	–	–	–
Trans-p-menth-2-en-1-ol, 1141, 1141	–	–	–	–	–	–	–	–	–	0.07	0.2	0.2
Trans-verbenol	–	–	–	–	–	–	–	–	0.12	–	–	–
Camphor	–	2.03	14.06	–	–	–	–	–	–	–	0.4	0.7
Menthone	–	0.21	–	–	–	–	–	–	–	–	–	–
Isoborneol	1.75	–	–	–	–	–	–	–	–	–	–	–
δ-terpineol	–	–	–	–	–	–	–	–	–	–	0.06	–
Nonanol	–	–	–	–	–	–	–	–	–	tr	–	–
Borneol	–	–	3.6	2.6	0.5	tr	1.9	3.4	0.3	6.3	4.4	8.2
Endo-borneol	–	0.26	–	–	–	–	–	–	–	–	–	–
Terpinen-4-ol	0.20	0.27	4.8	3.7	1.9	2.4	3.7	4.3	1.01	2.6	–	4.7
p-cymen-8-ol	1.63	–	–	–	–	–	–	–	0.08	–	–	–
α-terpineol	–	–	–	–	–	–	–	–	0.41	–	1.5	2.0
Cis-dihydrocarvone	0.10	–	–	–	–	–	–	–	–	0.2	–	–
Trans-dihydrocarvone	0.04	–	–	–	–	–	–	–	–	0.07	0.1	–
Trans-carveol	0.02	–	–	–	–	–	–	–	–	0.08	0.1	–
β-cyclocitral	–	–	–	–	–	–	–	–	–	0.07	–	–
Nerol	–	–	–	–	–	–	–	–	–	–	0.07	–
Thymol methyl ether	–	1.94	–	–	–	–	–	–	14.44	–	–	–
Carvacrol methyl ether	0.15	–	–	–	–	–	–	–	tr	–	–	–
Thymoquinone	–	–	–	–	–	–	–	–	0.10	–	–	–

Table A.1 (continued)

Plants name	*S. thymbra*								*S. spicigera*	*S. coerulea*		
Compounds name	Greece [152]	Turkey [99]	Turkey, Antalya (%v/w) [47]						Turkey [111]	Turkey [110]		
			Feb	March	April	May	June	July		July	Aug	Aug
Decanol	–	–	–	–	–	–	–	–	–	0.2	–	–
(E)-dec-2-enal	–	–	–	–	–	–	–	–	–	–	0.1	–
Nonanoic acid	–	–	–	–	–	–	–	–	–	0.2	–	–
Bornyl acetate	–	–	–	–	–	–	–	–	–	0.4	–	–
Thymol	34.72	13.19	96.6	92.2	169.1	172.3	366.3	418.2	23.99	0.8	7.9	–
Carvacrol	3.12	17.50	47.2	21.6	23.7	19.8	63.9	55.0	10.76	3.6	0.9	–
α-terpinyl acetate	–	–	–	–	–	–	–	–	0.24	–	–	–
α-cubebene	–	–	–	–	–	–	–	–	–	0.1	–	–
α-bourbonene	–	–	–	–	–	–	–	–	–	0.1	–	–
α-copaene	–	–	–	–	–	–	–	–	0.08	0.7	0.4	
decanoic acid	–	–	–	–	–	–	–	–	–	0.2	–	–
β-cubebene	–	–	–	–	–	–	–	–	–	1.1	–	–
β-bourbonene	–	–	–	–	–	–	–	–	0.17	2.3	4.4	0.3
β-elemene	–	–	–	–	–	–	–	–	–	4.2	4.9	0.3
β-damascone	–	–	–	–	–	–	–	–	–	–	0.08	–
α-humulene	–	0.19	–	–	–	–	–	–	0.23	1.3	1.4	0.5
(E)-geranyl acetone	–	–	–	–	–	–	–	–	–	0.3	0.1	–
β-caryophyllene	3.82	3.15	14.7	14.3	20.9	15.7	41.4	21.4	4.37	–	10.3	12.2
γ-muurolene	–	–	–	–	–	–	–	–	0.21	2.0	–	–
Germacrene D	–	–	–	–	–	–	–	–	0.47	20.6	15.5	12.8
β-ionone	–	–	–	–	–	–	–	–	–	0.2	–	–
β-selinene	–	–	–	–	–	–	–	–	–	0.4	–	–
Cadina-1,4-diene	–	–	–	–	–	–	–	–	–	0.06	–	–

Table A.1 (continued)

Plants name	*S. thymbra*								*S. spicigera*	*S. coerulea*		
Compounds name	Greece [152]	Turkey [99]	Turkey, Antalya (%v/w) [47]						Turkey [111]	Turkey [110]		
			Feb	March	April	May	June	July		July	Aug	Aug
α-selinene	–	–	–	–	–	–	–	–	–	1.2	–	–
Epi-cubebol	–	–	–	–	–	–	–	–	–	0.7	0.9	0.4
Bicyclogermacrene	–	–	–	–	–	–	–	–	–	1.8	4.0	0.9
α-muurolene	–	–	–	–	–	–	–	–	–	4.0	–	–
Guaia-3,7-diene	–	–	–	–	–	–	–	–	–	0.2	0.08	–
β-bisabolene	0.12	–	–	–	–	–	–	–	1.38	–	0.02	3.3
(E, E)-α-farnesene	–	–	–	–	–	–	–	–	–	0.1	–	–
γ-cadinene	–	–	–	–	–	–	–	–	0.57	1.1	tr	0.3
Cubenol	–	–	–	–	–	–	–	–	–	0.4	0.3	–
Calamenene	0.07	–	–	–	–	–	–	–	–	0.2	0.2	–
α-cadinene	–	–	–	–	–	–	–	–	–	2.0	2.2	0.5
(E)-nerolidol	–	–	–	–	–	–	–	–	–	–	–	0.7
Germacrene-D-4-ol	–	–	–	–	–	–	–	–	–	3.4	3.0	1.7
Spathulenol	0.50	–	–	–	–	–	–	–	0.45	1.1	3.2	3.6
Isocaryophylene oxide	–	–	–	–	–	–	–	–	–	0.3	0.5	0.2
Caryophyllene oxide	–	0.20	–	–	–	–	–	–	0.51	–	–	–
Caryophylladienol	–	–	–	–	–	–	–	–	–	0.5	1.1	0.4
Caryophyllenol II	–	–	–	–	–	–	–	–	–	1.0	1.1	–
Viridiflorene	–	–	–	–	–	–	–	–	0.24	–	–	–
β-oplopenone	–	–	–	–	–	–	–	–	–	0.4	0.4	–
1-epi-cubenol	–	–	–	–	–	–	–	–	–	0.3	0.1	tr
T-cadinol	–	–	–	–	–	–	–	–	–	1.6	1.0	–

Table A.1 (continued)

Plants name	*S. thymbra*								*S. spicigera*	*S. coerulea*		
Compounds name	Greece [152]	Turkey [99]	Turkey, Antalya (%v/w) [47]						Turkey [111]	Turkey [110]		
			Feb	March	April	May	June	July		July	Aug	Aug
T-muurolol	–	–	–	–	–	–	–	–	–	1.3	0.8	0.9
δ-cadinol	–	–	–	–	–	–	–	–	–	–	0.3	0.2
α-cadinol	–	–	–	–	–	–	–	–	–	3.8	2.5	2.3
Trans-α-bergamotol	–	–	–	–	–	–	–	–	–	–	0.4	–
Tetradecanoic acid	–	–	–	–	–	–	–	–	–	–	0.1	–
Benzyl benzoate	–	–	–	–	–	–	–	–	–	–	0.1	–
Hexahydrofarnesyl acetone	–	–	–	–	–	–	–	–	–	0.3	–	–
Hexadecanol	–	–	–	–	–	–	–	–	–	2.5	4.5	–
Phytol	–	–	–	–	–	–	–	–	–	0.1	0.07	–
Hexadecanoic acid	–	–	–	–	–	–	–	–	–	–	1.5	0.6
Tricosane	–	–	–	–	–	–	–	–	–	0.1	–	–
2,6-dimethyl-3-(E),5(E),7-octatriene-2-ol	–	–	–	–	–	–	–	–	–	0.1	0.2	0.4

tr trace

Table A.1 (continued)

Plants name / Compounds name	*S. punctata* ssp. *punctata* [102]	*S. parnassica* ssp. *parnassica* [106]		*S. isophylla* [108]	*S. rechingeri* [45]			*S. khuzestanica* [45]	*S. boissieri* [65, 73]		*S. coerulea* [73]	*S. icarica*	*S. pilosa*
		Flower	Vegetative		HD	HSD	SD						
Tricyclene 927	–	tr	0.49	–	–	–	–	–	–	–	–	–	–
α-thujene, 930	–	tr	0.73	0.7	0.9	–	–	–	1.4	–	–	–	–
α-pinene, 939	0.10	0.58	3.16	3.3	0.4	0.6	0.8	0.28	1.0	–	–	–	–
Camphene, 954	–	tr	1.96	4.4	–	–	0.1	–	0.1	–	–	–	–
Sabinene, 975	1.01	–	–	1.8	–	–	–	–	–	–	–	–	–
β-pinene, 979	–	tr	1.08	–	0.1	0.1	0.2	–	0.2	–	–	–	–
1-octen-3-ol, 979	–	–	–	–	–	–	–	–	–	–	–	–	–
Myrcene, 991	0.31	–	–	–	1.1	0.9	tr	0.39	2.8	–	–	–	–
3-octanol, 991	–	tr	tr	–	–	–	–	–	0.2	–	–	–	–
Yomogi alcohol 999	0.31	–	–	–	–	–	–	–	–	–	–	–	–
α-phellandrene, 1003	–	–	–	–	0.2	0.1	1.3	–	0.4	–	–	–	–
α-terpinene, 1017	–	–	–	–	0.5	0.4	0.1	0.49	4.6	–	–	–	–
6-methylhept-5-en-2-one	0.34	–	–	–	–	–	–	–	–	–	–	–	–
p-cymene, 1025	0.21	12.96	15.30	3.7	2.4	1.9	0.6	3.11	14.5	35.5	–	15.7	8.1
Limonene, 1029	0.36	0.56	1.00	3.9	0.2	0.2	2.6	–	–	–	4.3	–	–
β-phellandrene, 1030	0.10	–	–	–	–	–		0.19	–	–	–	–	–
1,8–cineole, 1031	–	0.37	0.94	–	–	–	0.3	–	–	–	–	–	–
δ-3-carene, 1031	–	–	–	–	0.1	0.1	0.3	–	0.1	–	–	–	–
Cis-β-ocimene, 1037	–	0.39	tr	–	–	–	–	–	–	–	–	–	–
β-ocimene	–	–	–	–	–	–	–	–	0.1	–	–	–	–
Trans-β-ocimene, 1050	0.11	tr	0.23	–	tr	1.5	0.1	–	–	–	–	–	–
γ-terpinene, 1060	–	0.90	1.70	2.0	–	–	–	1.24	26.4	6.5	–	4.4	–
n-octanol 1068	–	–	–	–	0.7		1.0	–	–	–	–	–	–
β-thujone	–	–	–	–	0.1	0.1	2.2	–	–	–	–	–	–
Cis-sabinene hydrate, 1070	0.65	0.58	3.52	–	–	–	–	–	–	–	–	–	–
α-terpinolene, 1089	–	tr	0.06	0.5	2.2	tr	2.3	–	–	–	–	–	–
p-cymenene 1091	–	tr	tr	–	–	–	–	–	–	–	–	–	–

Table A.1 (continued)

Plants name	*S. punctata* ssp. *punctata* [102]	*S. parnassica* ssp. *parnassica* [106]		*S. isophylla* [108]	*S. rechingeri* [45]			*S. khuzestanica* [45]	*S. boissieri* [65, 73]		*S. coerulea*	*S. icarica*	*S. pilosa*
Compounds name		Flower	Vegetative		HD	HSD	SD				[73]		
Linalool, 1097	0.91	2.56	12.85	–	–	–	–	0.91	1.3	–	–	–	–
Trans-sabinene hydrate, 1098	0.21	–	–	–	–	–	–	–	0.2	–	–	–	–
Nonanal, 1101	–	tr	–	–	–	–	–	–	–	–	–	–	–
Cis-rose oxide 1108	0.04	–	–	–	–	–	–	–	–	–	–	–	–
Thujone 1102 1114	0.04	–	–	–	–	–	–	–	–	–	–	–	–
1,3,8-*p*-menthatriene, 1110	–	tr	–	–	–	–	–	–	–	–	–	–	–
Trans-rose oxide 1126	0.27	–	–	–	–	–	–	–	–	–	–	–	–
Allo-ocimene, 1132	–	–	–	–	–	–	–	–	–	–	–	–	–
Veratrole, 1146	0.87	–	–	–	–	–	–	–	–	–	–	–	–
Camphor, 1146	–	tr	0.59	9.4	–	–	–	–	–	–	–	–	–
Citronellal, 1153	0.28	–	–	–	–	–	–	–	–	–	–	–	–
Geraniol, 1153	0.65	–	–	–	–	–	–	–	–	–	–	–	7.6
β-terpineol, 1163 1144	0.98	–	–	–	–	–	–	–	–	–	–	–	–
Cis-chrysanthenol, 1164	0.27	–	–	–	–	–	–	–	–	–	–	–	–
Geranial, 1167	27.65	–	–	–	–	–	–	–	–	–	–	–	–
Borneol, 1169	–	0.58	2.79	–	0.7	tr	0.1	0.35	0.1		6.3	4.5	4.7
p-mentha-1,5-diene-8-ol, 1170	0.51	–	–	–	–	–	–	–	–	–	–	–	–
Isoborneol, 1162	–	–	1.45	–	–	–	–	–	–	–	–	–	–
Terpinen-4-ol, 1177	–	0.44	1.88	2.3	–	tr	0.5	0.65	–	–	–	–	–
Cryptone, 1186	0.34	–	–	–	–	–	–	–	–	–	–	–	–
α-copaene, 1377	–	–	–	0.2	–	–	–	–	–	–	–	–	–
Cis-dihydrocarvone, 1193	–	tr	tr	–	–	–	–	–	–	–	–	–	–
α-terpineol, 1198	0.35	tr	0.65	0.4	–	–	–	–	tr	–	–	–	–
Dehydrolinalool	0.29	–	–	–	–	–	–	–	–	–	–	–	–
Nerol, 1235	–	tr	tr	–	–	–	–	–	–	–	–	–	–
Neral, 1238	21.72	tr	tr	–	0.2	0.2	0.1	–	–	–	–	–	–

Table A.1 (continued)

Plants name	S. punctata ssp. punctata [102]	S. parnassica ssp. parnassica [106]		S. isophylla [108]	S. rechingeri [45]			S. khuzestanica [45]	S. boissieri [65, 73]		S. coerulea [73]	S. icarica	S. pilosa
Compounds name		Flower	Vegetative		HD	HSD	SD						
β-cedrene, 1241	0.11	–	–	–	–	–	–	–	–	–	–	–	–
Cuminaldehyde, 1242	0.18	–	–	–	–	–	–	–	–	–	–	–	–
Carvacrol methyl ether, 1245	–	–	–	–	–	–	–	–	0.2	–	–	2.0	3.1
Piperitone, 1253	0.32	–	–	–	–	–	–	–	–	–	–	–	–
p-mentha-1-en-7-al, 1276	0.33	–	–	–	–	–	–	–	–	–	–	–	–
Dihydro edulan II, 1284	0.34	–	–	–	–	–	–	–	–	–	–	–	–
Bornyl acetate, 1289	–	–	–	–	0.1		0.2	–	–	–	–	–	–
Thymol, 1290	–	7.45	8.50	–	–	–	–	0.19	–	2.3	–	59.2	42.1
Carvacrol, 1299	–	20.4	1.59	–	86.6	89.3	84.0	90.88	–	44.8	–	–	–
Ethyl nerolate	0.32	–	–	–	–	–	–	–	–	–	–	–	–
Neryl acetate, 1362	0.10	–	–	–	–	–	–	–	–	–	–	–	–
α-ylangene, 1375	–	tr	tr	–	–	–	–	–	40.8	–	–	–	–
Geranyl acetate, 1381	0.67	–	–	–	–	–	–	–	–	–	–	–	–
β-bourbonene, 1388	0.87	tr	tr	4.5	–	–	–	–	–	–	3.6	–	–
Cis-caryophyllene, 1409	–	tr	tr	–	–	–	–	–	–	–	–	–	–
α-gurjunene 1410	–	–	0.75	–	–	–	–	–	–	–	–	–	–
Trans-caryophyllene, 1419	–	20.85	7.13	–	–	–	–	–	–	–	–	–	–
p-cuminyl acetate, 1419	0.17	–	–	–	–	–	–	–	–	–	–	–	–
β-gurjunene, 1434	–	tr	–	0.6	–	–	–	–	–	–	–	–	–
α-trans-bergamotene, 1435	–	tr	tr	–	–	–	–	–	–	–	–	–	–
Aromadendrene, 1441	0.12	tr	tr	–	–	–	–	–	0.2	–	–	–	–
α-humulene, 1455	0.16	0.79	tr	–	–	–	–	–	tr	–	–	–	–
β-caryophyllene, 1466	0.72	–	–	12.1	–	–	–	–	0.8	–	106	–	–
γ-muurolene, 1480	–	tr	tr	0.3	–	–	–	–	–	–	–	–	–
Germacrene D, 1485	1.43	tr	tr	2.1	–	–	–	–	–	–	4.7	–	–
β-selinene, 1490	–			0.2	–	–	–	–	–	–	–	–	–
Bicyclogermacrene, 1500	–	tr	0.81	1.0	–	–	–	–	–	–	–	–	–
α-muurolene, 1500	–	tr	tr	0.2	–	–	–	–	–	–	–	–	–

Table A.1 (continued)

Plants name	*S. punctata* ssp. *punctata* [102]	*S. parnassica* ssp. *parnassica* [106]		*S. isophylla* [108]	*S. rechingeri* [45]			*S. khuzestanica* [45]	*S. boissieri* [65, 73]		*S. coerulea* [73]	*S. icarica*	*S. pilosa*
Compounds name		Flower	Vegetative		HD	HSD	SD						
β-bisabolene, 1506	2.78	5.52	3.36	–	0.5	0.7	1.3	0.21	–	–	–	–	–
γ-cadinene 1514	–	tr	tr	2.4	–	–	–	–	tr	–	–	–	–
δ-cadinene, 1523	–	tr	0.66	–	–	–	–	–	–	–	–	–	–
α-cadinene, 1539	–	–	tr	–	–	–	–	–	–	–	–	–	–
(Z)-β-sesquisabinene hydrate, 1544	–	–	–	–	0.1	–	0.3	–	–	–	–	–	–
α-calacorene, 1546	–	–	tr	–	–	–	–	–	–	–	–	–	–
Elemol, 1550	–	–	–	4.7	–	–	–	–	–	–	–	–	–
Spathulenol, 1578	–	17.18	tr	0.8	–	–	–	–	0.2	–	3.0	–	–
Caryophylene oxide, 1583	2.70	1.70	16.34	1.7	–	–	–	0.18	0.1	–	–	–	–
Globulol 1585	1.09	–	–	–	–	–	–	–	–	–	–	–	–
Viridiflorol 1593	–	–	–	–	–	–	–	–	tr	–	–	–	–
Ledene (viridiflorene 1593)	–	–	–	–	–	–	–	–	0.2	–	–	–	–
Humulene epoxide II 1608	–	tr	0.61	–	–	–	–	–	–	–	–	–	–
Humulene epoxide I	0.37	–	–	–	–	–	–	–	–	–	–	–	–
caryophylla-4(14),8(15)-dien-5α-ol 1641	–	tr	tr	–	–	–	–	–	–	–	–	–	–
Caryophylla-4(14),8(15)-dien-5β-ol, 1641	–	1.94	1.29	–	–	–	–	–	–	–	–	–	–
Trans-nerolidol	4.82	–	–	–	–	–	–	–	–	–	–	–	–
α-eudesmol 1654	–	–	–	24.2	–	–	–	–	–	–	–	–	–
14-hydroxy-9-epi-(E)-caryophyllene 1670	–	1.33	2.09	–	–	–	–	–	–	–	–	–	–
Lanceol, 1761	–	0.26	–	–	–	–	–	–	–	–	–	–	–
Hexadecanoic acid, 1984	–	–	–	–	–	–	–	–	tr	–	–	–	–

tr trace

Table A.1 (continued)

Plants name	*S. parvifolia* [95, 96]		*S. boliviana* [2, 95]		*S. sahendica* flowering period [134]			*S. sahendica* different altitudes and regions [91]							
Compounds name					Before	During	After	S1	S2	S3	S4	S5	S6	S7	S8
Tricyclene 927	–	–	0.7	–	–	–	–	–	–	–	–	–	–	–	–
α-thujene, 930	–	0.1	0.1	0.3	0.93	0.87	0.89	0.9	0.6	0.5	0.8	tr	0.9	0.8	3.3
α-pinene, 939	–	0.3	0.6	0.3	0.63	0.53	0.48	0.6	0.7	0.5	0.6	0.4	0.6	0.6	0.6
Camphene, 954	–	tr	1.4	–	0.15	0.22	0.18	0.1	tr	tr	tr	–	tr	0.1	0.9
Sabinene, 975	–	–	0.8	0.6	0.1	0.08	0.07	–	–	–	–	–	–	–	–
β-pinene, 979	–	0.5	0.4	–	0.25	0.33	0.28	0.1	1.0	0.3	tr	0.7	0.2	0.2	0.3
1-octen-3-ol, 979	–	–	0.6	tr	–	–	–	–	–	–	–	–	–	–	–
3-octanol, 991	–	tr	–	tr	–	–	–	–	–	–	–	–	–	–	–
Myrcene, 991	–	0.9	0.2	0.4	2.28	2.12	1.87	1.5	1.1	1.3	1.4	tr	1.5	1.4	1.0
α-phellandrene, 1003	–	–	–	0.1	0.28	0.27	0.33	0.1	0.3	0.2	tr	0.2	0.2	0.2	–
α-terpinene, 1017	–	0.4	–	0.1	1.75	1.43	2.12	1.4	1.1	1.0	1.3	tr	1.4	1.2	0.6
o-cymene, 1011	–	–	1.4	–	–	–	–	–	–	–	–	–	–	–	–
p-cymene, 1025	–	0.1	2.8	3.4	23.89	34.33	30.28	32.5	33.0	47.1	41.7	33.4	33.2	37.4	54.9
Limonene, 1029	6.0	0.9	0.6	–	1.54	1.33	1.63	0.5	2.0	1.1	0.5	2.1	0.8	0.5	1.4
β-phellandrene, 1030	–	–	–	–	0.14	0.18	0.22	–	–	–	–	–	–	–	–
1,8-cineole, 1031	–	1.2	7.4	4.0	0.13	0.11	0.09	tr	0.2	tr	tr	tr	tr	tr	–
δ-3-carene, 1031	–	–	–	–	0.08	0.14	0.09	tr	tr	–	tr	–	tr	tr	–
Cis-β-ocimene, 1037	–	0.1	–	0.1	–	–	–	–	–	–	–	–	–	–	–
Trans-β-ocimene, 1050	–	0.4	3.4	0.1	0.06	0.05	0.07	tr	tr	–	tr	–	tr	tr	–
γ-terpinene, 1060	–	–	15.4	0.8	13.78	17.75	19.37	10.9	9.1	8.9	11.0	1.0	12.8	9.9	3.2
Cis-sabinene hydrate, 1070	–	tr	–	–	–	–	–	–	–	–	–	–	–	–	–
m-cymenene 1085	–	tr	–	–	–	–	–	–	–	–	–	–	–	–	–
Cis-linalool oxide 1087	–	tr	–	–	–	–	–	–	–	–	–	–	–	–	–
α-terpinolene, 1089	–	–	–	0.1	0.23	0.18	0.15	0.2	0.1	–	0.3	–	–	tr	0.3

Table A.1 (continued)

Plants name	*S. parvifolia* [95, 96]		*S. boliviana* [2, 95]		*S. sahendica* flowering period [134]			*S. sahendica* different altitudes and regions [91]							
Compounds name					Before	During	After	S1	S2	S3	S4	S5	S6	S7	S8
Linalool, 1097	–	1.0	4.8	3.1	1.32	1.13	1.23	1.0	0.9	1.7	1.2	0.4	1.0	0.8	1.4
Trans-sabinene hydrate, 1098	–	–	–	–	–	–	–	0.2	0.2	0.2	0.3	tr	0.2	0.2	0.2
Trans-thujone 1114	–	0.1	–	–	–	–	–	–	–	–	–	–	–	–	–
Cis-p-menth-2-en-1-ol, 1122	–	–	–	0.1	–	–	–	–	–	–	–	–	–	–	–
Dihydrolinalool 1135	–	0.1	–	–	–	–	–	–	–	–	–	–	–	–	–
Trans-p-menth-2-en-1-ol, 1141	–	–	tr	–	0.03	0.05	0.08	tr	tr	tr	tr	tr	tr	tr	–
Trans-verbenol, 1145	–	0.1	–	–	–	–	–	–	–	–	–	–	–	–	–
Camphor, 1146	–	–	–	–	0.09	0.13	0.15	tr	0.2	tr	tr	tr	tr	tr	–
Isopulegol 1150	–	tr	–	–	–	–	–	–	–	–	–	–	–	–	–
Menthone, 1153	7.0	–	–	24.2	–	–	–	–	–	–	–	–	–	–	–
Geraniol, 1153	–	0.3	–	–	–	–	–	–	–	–	–	–	–	–	–
Isoborneol, 1162	–	tr	tr	–	–	–	–	–	–	–	–	–	–	–	–
Isomenthone, 1164	–	–	–	29.7	–	–	–	–	–	–	–	–	–	–	–
Borneol, 1169	–	–	–	–	0.05	0.21	0.11	tr	tr	tr	0.3	tr	–	0.1	0.6
Menthol, 1172	21.0	tr	–	–	–	–	–	–	–	–	–	–	–	–	–
Terpinen-4-ol, 1177	–	tr	0.6	–	–	–	–	0.6	0.4	tr	0.6	tr	0.6	0.5	0.7
p-cymen-8-ol, 1183	–	0.6	–	–	–	–	–	0.3	tr	tr	0.4	tr	tr	0.3	0.7
Dihydrocarvone, 1193 1201	–	–	–	6.6	–	–	–	–	2.2	0.9	–	2.8	0.4	tr	–
Estragole, 1195	–	–	–	–	–	–	–	–	–	0.7	–	0.5	tr		–
Myrtenol, 1196	–	0.6	–	–	–	–	–	–	–	–	–	–	–	–	–
α-terpineol, 1198	2.0	0.4	–	tr	–	–	–	tr	tr	tr	tr		tr	tr	–
Thuja-2,4(10)-diene	–	–	–	–	–	–	–	tr	tr	–	tr	–	tr	tr	–
Trans-pinan-2-ol	–	–	3.9	–	–	–	–	–	–	–	–	–	–	–	–
Decanal, 1204	4.0	–	–	–	–	–	–	–	–	–	–	–	–	–	–

Table A.1 (continued)

Plants name	*S. parvifolia* [95, 96]		*S. boliviana* [2, 95]		*S. sahendica* flowering period [134]			*S. sahendica* different altitudes and regions [91]							
Compounds name					Before	During	After	S1	S2	S3	S4	S5	S6	S7	S8
Verbenone, 1205	–	tr	–	–	–	–	–	–	–	–	–	–	–	–	–
Trans-carveol, 1217	–	tr	–	0.1	–	–	–	–	–	–	–	–	–	–	–
Citronellol, 1228	4.0	0.6	–	–	–	–	–	–	–	–	–	–	–	–	–
Pulegone, 1237	–	4.4	–	10.7	–	–	–	–	1.1	0.4	–	1.2	tr	–	0.3
Cuminaldehyde, 1242	–	0.6	–	0.4	–	–	–	tr	–	–	tr	–	–	tr	–
Cumin alcohol	–	–	–	0.6	–	–	–	–	–	–	–	–	–	–	–
Carvone, 1243	–	0.6	–	–	–	–	–	–	–	–	–	–	–	–	–
Piperitone oxide, 1251	31.0	–	–	–	–	–	–	–	–	–	–	–	–	–	–
Piperitone, 1253	1.9	0.8	–	2.1	–	–	–	–	–	–	–	–	–	–	–
Trans-myrtanol, 1261	–	0.8	–	–	–	–	–	–	–	–	–	–	–	–	–
Linalyl acetate 1257	–	–	0.4	0.1	–	–	–	–	–	–	–	–	–	–	–
Perilla aldehyde, 1272	–	–	–	0.1	–	–	–	–	–	–	–	–	–	–	–
Citronellyl formiate, 1274	–	1.4	–	–	–	–	–	–	–	–	–	–	–	–	–
Methyl nerolate, 1279	–	0.6	–	–	–	–	–	–	–	–	–	–	–	–	–
Trans-anethole, 1285	–	–	–	0.1	–	–	–	–	–	–	–	–	–	–	–
α-terpinen-7-al, 1285	–	–	–	–	–	–	–	1.0	0.7	0.9	1.4	1.6	1.0	0.7	–
Bornyl acetate, 1289	–	–	2.2	–	–	–	–	–	–	–	–	–	–	–	–
Thymol, 1290	–	tr	–	4.5	42.24	28.76	31.53	41.7	33.4	27.5	30.2	32.7	37.6	41.2	19.6
Pinocarvyl acetate, 1297	–	–	0.3	–	–	–	–	–	–	–	–	–	–	–	–
Carvacrol, 1299	–	–	–	0.1	1.75	0.83	1.12	1.1	0.8	0.7	1.3	0.9	1.0	0.8	1.2
Trans-carvyl acetate, 1342	–	tr	–	–	–	–	–	–	–	–	–	–	–	–	–
Piperitenone, 1343	–	5.6	–	–	–	–	–	–	–	–	–	–	–	–	–
δ-elemene, 1338	–	–	0.9	–	–	–	–	–	–	–	–	–	–	–	–
Thymol acetate, 1352	–	–	–	tr	0.87	1.12	0.75	tr	0.3	tr	0.4	8.2	1.2	tr	–

Table A.1 (continued)

Plants name	*S. parvifolia* [95, 96]		*S. boliviana* [2, 95]		*S. sahendica* flowering period [134]			*S. sahendica* different altitudes and regions [91]							
Compounds name					Before	During	After	S1	S2	S3	S4	S5	S6	S7	S8
Neryl acetate, 1362	–	–	0.1	–	–	–	–	–	–	–	–	–	–	–	–
Piperitenone oxide, 1369	16.0	69.8	–	–	–	–	–	–	–	–	–	–	–	–	–
α-copaene, 1377	–	–	0.2	–	–	–	–	–	–	–	–	–	–	–	–
β-cubebene, 1388	–	–	–	tr	–	–	–	–	–	–	–	–	–	–	–
β-bourbonene, 1388	–	–	0.1	0.1	–	–	–	–	–	–	–	–	–	–	–
1-tetradecyne, 1391	–	–	–	0.1	–	–	–	–	–	–	–	–	–	–	–
β-elemene, 1391	–	–	0.3	0.1	–	–	–	–	–	–	–	–	–	–	–
Longifolene, 1408	–	–	–	tr	–	–	–	–	–	–	–	–	–	–	–
α-gurjunene 1410	–	–	0.1	tr	–	–	–	–	–	–	–	–	–	–	–
Trans-caryophyllene, 1419	–	0.9	10.2	–	–	–	–	–	–	–	–	–	–	–	–
β-gurjunene, 1434	–	–	0.1	–	–	–	–	–	–	–	–	–	–	–	–
γ-elemene, 1437	–	–	–	1.3	–	–	–	–	–	–	–	–	–	–	–
2-phenyl ethyl butanoate, 1439	–	–	–	–	–	–	–	0.8	–	0.8	–	tr	1.3	–	–
α-guaiene, 1440	–	–	–	tr	–	–	–	–	–	–	–	–	–	–	–
Aromadendrene, 1441	–	–	0.2	0.1	0.12	0.26	0.14	–	–	–	–	–	–	–	–
Cis-β-farnesene, 1443	–	0.1	–	–	–	–	–	–	–	–	–	–	–	–	–
α-humulene, 1455	–	–	–	0.1	0.13	0.08	0.14	tr	–	–	tr	–	–	tr	–
Alloaromadendrene, 1461	–	–	0.2	0.1	–	–	–	–	–	–	–	–	–	–	–
β-caryophyllene, 1466	–	–	–	1.6	1.84	2.26	1.83	2.0	1.7	0.7	2.0	0.6	1.6	0.5	0.9
Cis-muurola-4(14),5-diene, 1467	–	–	0.1	–	–	–	–	–	–	–	–	–	–	–	–
Cis-muurola-4(14),5-diene, 1467	–	–	0.1	–	–	–	–	–	–	–	–	–	–	–	–
γ-gurjunene, 1477	–	–	–	tr	–	–	–	–	–	–	–	–	–	–	–
γ-muurolene, 1480	–	tr	0.5	0.2	–	–	–	–	–	–	–	–	–	–	–
Germacrene D, 1485	–	–	8.9	–	–	–	–	–	–	–	–	–	–	–	–

Table A.1 (continued)

Plants name	*S. parvifolia* [95, 96]		*S. boliviana* [2, 95]		*S. sahendica* flowering period [134]			*S. sahendica* different altitudes and regions [91]							
Compounds name					Before	During	After	S1	S2	S3	S4	S5	S6	S7	S8
β-selinene, 1490	–	tr	–	tr	–	–	–	–	–	–	–	–	–	–	–
Valencene, 1496	–	–	–	–	0.08	0.12	0.09	tr	–	–	–	–	–	tr	–
α-muurolene, 1500	–	–	–	0.2	–	–	–	–	–	–	–	–	–	–	–
β-bisabolene, 1506	–	–	–	–	–	–	–	tr	3.3	1.4	–	4.5	0.9	tr	0.8
Trans-α-bisabolene, 1507	–	–	–	–	1.85	1.75	1.63	–	–	–	–	–	–	–	–
γ-cadinene 1514	–	–	–	0.3	–	–	–	–	–	–	–	–	–	–	–
δ-cadinene, 1523	–	–	–	0.7	–	–	–	–	–	–	–	–	–	–	–
α-cadinene, 1539	–	–	–	tr	–	–	–	–	–	–	–	–	–	–	–
Elemol, 1550	–	–	–	0.1	–	–	–	–	–	–	–	–	–	–	–
Spathulenol, 1578	3.0	–	–	1.3	0.32	0.42	0.29	0.8	0.7	0.3	1.1	1.3	0.6	0.7	0.9
Caryophylene oxide, 1583	–	–	–	tr	0.87	1.14	0.75	1.2	1.6	0.7	2.5	1.6	1.2	0.4	2.5
Viridiflorol 1593	–	–	–	–	–	–	–	–	–	–	tr	–	–	–	–
Guaiol, 1601	–	–	–	0.1	–	–	–	–	–	–	–	–	–	–	–
γ-cadinol, 1640	–	–	–	0.1	–	–	–	–	–	–	–	–	–	–	–
α-cadinol 1654	–	–	–	0.1	–	–	–	–	–	–	–	–	–	–	–
olopanone, 1740	–	–	–	0.3	–	–	–	–	–	–	–	–	–	–	–

S1, S2, S3 Iran, Zanjan, road towards Dandi, Ghezlu, 2315 m, Jamzad et al. 83143 (TARI), 83142 (TARI), 83141 (TARI), respectively, *S4* Iran, Zanjan, S. of Soltanieh, shalvar village, 2047 m, Jamzad et al. 83144 (TARI), *S5, S6* Iran, Zanjan, Gheidar, Zarrin abad, Gholamveis village, 2105 m, Jamzad et al. 83146 (TARI), 83146 (TARI), respectively, *S7* Iran, Azarbayejan, Miandoab, 2400–2600 m, Ghahramani, s.n. *S8* Iran, Kurdestan, 25 Km from Bijar to Takab, Salavat–abad village, 1950 m, Maroofi and Naser, 5552. *tr* trace

Table A.1 (continued)

Plants name	*S. bachtarica* [66]			*S. cuneifolia* [75]	*S. subspicata* [160]		*S. darwinii* [34]	*S. multiflora* [34]	*S. punctata* ssp. *punctata* [102]	*S. biflora* [93]	*S. masukensis* [93]	*S. pseudosimensis* [93]
Compounds name	Leaf	Stem	Aerial part		Summer	Autumn						
Hexanal, 802	–	–	–	–	–	–	–	–	–	0.03	–	–
Cis-3-hexenol, 859	–	–	–	–	–	–	–	–	–	0.04	–	0.13
1-hexanol, 871	–	–	–	–	–	–	–	–	–	0.03	–	–
Hexanoic acid, 890	–	–	–	–	–	–	0.5	–	–	–	–	–
6-methylheptanol	–	–	–	–	–	–	0.6	–	–	–	–	–
2,6-octadienal	–	–	–	0.83	–	–	–	–	–	–	–	–
Tricyclene 927	–	–	–	–	–	–	–	–	–	–	0.02	0.01
α-thujene, 930	0.03	0.03	0.15	–	–	–	–	–	–	0.03	0.02	0.01
α-pinene, 939	0.03	0.5	0.2	0.6	–	–	–	–	0.1	0.33	0.21	0.31
3-methylcyclohexanone, 952	–	–	–	–	–	–	0.1	–	–	–	–	–
Camphene, 954	–	–	–	0.57	–	–	–	–	–	0.22	0.23	0.34
Benzaldehyde, 960	–	–	–	–	–	–	–	–	–	0.09	0.03	0.04
p-mentha-1(7),8-diene, 960	–	–	–	–	–	–	–	–	–	–	0.04	–
Sabinene, 975	–	–	–	–	–	–	–	–	1.01	0.05	0.06	0.10
1-octen-3-ol, 979	–	–	–	–	–	–	–	–	0.21	2.82	0.02	0.14
β-pinene, 979	–	–	–	–	–	–	–	–	–	0.14	0.07	0.30
3-octanone 984	–	–	–	–	–	–	–	–	–	0.66	–	0.13
6-methyl-5-hepten-2-one, 986	–	–	–	–	–	–	–	–	–	–	0.06	–
Myrcene, 991	0.05	–	0.58	0.53	–	–	–	–	0.31	0.06	0.14	0.29
3-octanol, 991	–	–	–	–	–	–	0.7	–	–	–	–	–
3-octanal, 999	–	–	–	–	–	–	–	–	–	0.98	0.37	–
α-phellandrene, 1003	0.03	–	0.08	0.69	–	–	–	–	–	0.03	–	0.03
α-terpinene, 1017	0.05	0.18	0.47	0.97	–	–	–	–	–	0.05	–	–

Table A.1 (continued)

Plants name	*S. bachtarica* [66]			*S. cuneifolia* [75]	*S. subspicata* [160]		*S. darwinii* [34]	*S. multiflora* [34]	*S. punctata* ssp. *punctata* [102]	*S. biflora* [93]	*S. masukensis* [93]	*S. pseudosimensis* [93]
Compounds name	Leaf	Stem	Aerial part		Summer	Autumn						
p-cymene, 1025	0.65	1.66	1.85	21.61	–	–	–	–	–	1.98	0.36	0.02
Limonene, 1029	–	–	–	–	–	–	–	–	0.31	0.55	0.45	2.62
β-phellandrene, 1030	–	0.05	–	–	–	–	–	–	–	–	–	–
1,8-cineole, 1031	–	–	–	1.66	–	–	–	–	0.21	1.92	–	–
δ-3-carene, 1031	–	–	–	–	–	–	–	27.65	0.34	–	–	–
1,8-cineol, 1031	0.05	1.71	0.10	–	–	–	–	–	–	–	–	–
α-ocimene, 1039	–	–	–	0.37	–	–	–	–	–	–	–	–
Cis-β-ocimene, 1037	–	–	–	–	–	–	–	–	0.11	–	–	–
β-ocimene 1050 1037	–	–	–	0.23	–	–	–	–	0.36	0.29	0.38	0.39
Phenylacetaldehyde, 1043	–	–	–	–	–	–	1.8	0.3	–	–	–	–
Trans-β-ocimene, 1050	–	–	–	–	–	–	–	–	0.10	–	–	–
γ-terpinene, 1060	0.23	1.05	1.18	4.35	–	–	–	–	–	0.34	0.02	–
Cis-sabinene hydrate, 1070	0.06	–	–	–	–	–	–	–	–	–	–	–
Trans-linalool oxide, 1073	–	–	–	–	–	–	0.4	–	0.65	3.15	0.69	0.16
Cis-linalool oxide 1087	–	–	–	–	–	–	0.4	–	0.29	–	–	–
Dehydrolinalool	–	–	–	–	–	–	–	–	0.91	–	–	–
Linalool, 1097	2.08	5.10	2.56	0.97	–	–	1.3	3.7	0.04	6.03	4.44	0.25
Trans-sabinene hydrate, 1098	0.06	–	–	–	–	–	–	–	–	0.61	–	–
Nonanal, 1101	–	–	–	–	–	–	–	–	0.27	–	–	–
β-thujone	–	–	–	–	–	–	–	–	–	0.29	–	–
Thuja-2,4(10)-diene	–	–	–	–	–	–	–	–	–	–	–	0.02
Cis-rose oxide, 1108	–	–	–	–	–	–	–	–	0.28	–	–	–
1-octen-3-yl acetate 1113	–	–	–	–	–	–	–	–	–	–	0.06	1.36
Cis-p-menth-2-en-1-ol, 1122	–	–	–	–	–	–	–	–	–	0.87	–	–

Table A.1 (continued)

Plants name	*S. bachtarica* [66]			*S. cuneifolia* [75]	*S. subspicata* [160]		*S. darwinii* [34]	*S. multiflora* [34]	*S. punctata* ssp. *punctata* [102]	*S. biflora* [93]	*S. masukensis* [93]	*S. pseudosimensis* [93]
Compounds name	Leaf	Stem	Aerial part		Summer	Autumn						
Trans-*p*-menth-2,8-dien-1-ol, 1123	–	–	–	–	–	–	0.6	–	0.04	–	–	–
3-octanol acetate 1123	–	–	–	–	–	–	–	–	–	–	0.64	2.79
α-campholenal, 1126	–	–	–	–	–	–	–	–	–	0.25	–	–
Trans-rose oxide, 1126	–	–	–	–	–	–	–	–	0.27	–	–	–
Cis-limonene oxide, 1138	–	0.03	–	–	–	–	–	–	–	–	–	–
Trans-pinocarveol, 1139	–	–	0.06	–	–	–	–	–	0.33	1.50	–	–
Trans-p-menth-2-en-1-ol, 1141	–	–	–	–	–	–	–	–	0.87	–	–	–
Verbenol, 1141	–	–	–	–	–	–	–	–	0.34		–	–
β-terpineol, 1163 1144	–	–	–	–	–	–	–	–	0.65	–	–	–
Camphor, 1146	–	0.03	–	–	–	–	–	–	–	3.89	0.49	0.12
Cis-chrysanthenol, 1146	–	–	–	–	–	–	–	–	0.98	–	–	–
Menthone, 1153	–	–	–	–	–	–	–	1.9	0.51	0.27	0.35	2.49
Geraniol, 1153	–	–	–	1.73	–	–		–	0.32		–	–
Sabina ketone, 1159	–	–	–	–	–	–	–	–	–	0.38	–	–
Endo-borneol, 1160	–	–	–	–	–	–	–	–	–	–	–	0.11
Pinocarvone, 1162	–	–	–	–	–	–	–	–	–	0.93	–	–
Isomenthone, 1164	–	–	–	–	–	–	–	83.1	0.34	–	–	8.47
Geranial, 1167	–	–	–	–	–	–	–	–	0.17	–	–	–
Borneol, 1169	0.74	0.83	0.44	2.51	–	–	2.7	–	0.10	2.97	2.44	–
Terpinen-4-ol, 1177	2.62	5.40	2.93	2.04	–	–	2.8	–	–	–	–	–
p-methyl-acetophenone, 1183	–	–	–	–	–	–	–	–	–	0.28	–	
p-cymen-8-ol, 1183	0.37	1.18	–	–	–	–	–	–	0.16	–	–	–

Table A.1 (continued)

Plants name	*S. bachtarica* [66]			*S. cuneifolia* [75]	*S. subspicata* [160]		*S. darwinii* [34]	*S. multiflora* [34]	*S. punctata* ssp. *punctata* [102]	*S. biflora* [93]	*S. masukensis* [93]	*S. pseudosimensis* [93]
Compounds name	Leaf	Stem	Aerial part		Summer	Autumn						
Cis-dihydrocarvone, 1193	–	–	–	–	–	–	–	–	2.78	–	–	–
α-terpineol, 1198	3.23	2.79	1.25	0.47	–	–	2.4	–	–	–	–	–
Trans-dihydrocarvone, 1201	–	1.20	–	–	–	–	–	–	–	–	–	–
Verbenone, 1205	–	–	–	–	–	–	1.3	–	–	–	–	–
Cis-*p*-menth-1(7),8-diene-2-ol, 1231	0.53		–	–	–	–	–	–	–	–	–	–
Thymol methyl ether, 1235	–	–	–	–	–	–	–	–	0.32	–	–	–
Nerol, 1235	–	–	–	0.15	–	–	–	–	–	–	–	–
Pulegone, 1237	–	–	–	–	–	–	11.4	0.7	4.82	–	–	–
Cis-carvyl acetate, 1242	–	–	–	–	–	–	–	–	–	–	–	0.08
Carvone, 1243	–	–	–	–	–	–	–	–	0.67	–	–	–
Carvacrol methyl ether, 1245	–	–	–	1.83	–	–	–	–	–	–	–	–
Piperitone oxide, 1251	–	–	–	–	–	–	–	–	–	–	0.07	0.24
Piperitone, 1253	–	–	–	–	–	–	2.1	1.3	–	–	–	–
Dihydro edulan II, 1284	–	–	–	–	–	–	–	–	0.72	–	–	–
Anethole, 1285	–	–	–	–	–	–	–	–	2.70	–	–	–
Bornyl acetate, 1289	–	–	–	–	–	–	–	–	–	4.75	1.84	–
Thymol, 1290	–	–	–	9.01	–	–	0.8	–	–	–	3.13	6.67
p-cymen-7-ol, 1291	–	–	–	–	–	–	–	–	–	0.47	–	–
Carvacrol, 1299	39.30	39.40	67.88	44.99	–	–	4.8	–	–	0.14	–	–
Trans-carvyl acetate, 1342	–	–	–	–	–	–	–	–	–	–	–	0.08
Piperitenone, 1343	–	–	–	–	–	–	57.8	–	0.37	0.09	0.09	0.51
α-terpinyl acetate, 1349	–	–	–	–	–	–	–	–	1.09	–	–	–

Table A.1 (continued)

Plants name	*S. bachtarica* [66]			*S. cuneifolia* [75]	*S. subspicata* [160]		*S. darwinii* [34]	*S. multiflora* [34]	*S. punctata* ssp. *punctata* [102]	*S. biflora* [93]	*S. masukensis* [93]	*S. pseudosimensis* [93]
Compounds name	Leaf	Stem	Aerial part		Summer	Autumn						
p-methoxy-acetophenone, 1350	–	–	–	–	–	–	–	–	–	–	–	0.05
Thymol acetate, 1352	0.36	–	–	–	–	–	–	–	–	–	–	–
α-copaene, 1377	–	–	–	–	–	–	–	–	–	0.39	0.71	0.25
Geranyl acetate, 1381	0.66	–	0.79	–	–	–	–	–	–	0.24	–	–
Trans-β-damascenone, 1385	0.13	0.09	–	–	–	–	–	–	–	–	–	–
β-bourbonene, 1388	–	–	–	–	–	–	–	–	0.11	4.19	5.53	2.93
β-cubebene, 1388	–	–	–	–	–	–	–	–	0.87	–	–	–
β-elemene, 1391	–	–	–	–	–	–	–	–	–	0.10	0.55	0.21
Trans-pinene hydrate	0.06	0.16	–	–	–	–	–	–	–	–	–	–
Cuminal	–	0.8	–	–	–	–	–	–	–	–	–	–
Eugenol, 1359	0.91	2.38	0.66	–	–	–	–	–	–	0.09	0.12	0.12
Cis-jasmone, 1391	–	–	–	–	–	–	–	–	–	–	0.24	
Methyl-eugenol, 1401	–	–	–	–	–	–	–	–	–	0.64	–	0.45
p-cuminyl acetate, 1419	–	–	–	–	–	–	–	–	0.12	–	–	–
Trans-caryophyllene, 1419	–	–	–	–	–	–	–	–	–	0.64	1.76	0.74
β-cedrene, 1421	–	–	–	–	–	–	–	–	0.18	–	–	–
β-gurjunene, 1434	–	–	–	–	–	–	–	–	0.35	0.15	1.02	0.38
Aromadendrene, 1441	–	–	–	–	–	–	–	–	21.72	–	0.11	0.16
α-humulene, 1455	0.09	–	0.08	–	–	–	–	–	–	–	0.16	0.34
Alloaromadendrene, 1461	–	–	–	–	–	–	–	–	–	0.52	–	–
β-caryophyllene, 1466	–	–	–	0.95	–	–	–	–	–	–	–	–
α-amorphene, 1485	–	–	–	0.25	–	–	–	–	–	0.25	0.53	0.26
β-ionone, 1489	0.39	0.59	0.26	–	–	–	–	–	–	0.42	0.57	0.32

Table A.1 (continued)

Plants name	*S. bachtarica* [66]			*S. cuneifolia* [75]	*S. subspicata* [160]		*S. darwinii* [34]	*S. multiflora* [34]	*S. punctata* ssp. *punctata* [102]	*S. biflora* [93]	*S. masukensis* [93]	*S. pseudosimensis* [93]
Compounds name	Leaf	Stem	Aerial part		Summer	Autumn						
α-selinene, 1498	–	–	–	–	–	–	–	–	–	0.03	–	–
Neoisomenthol	–	–	–	–	–	–	–	–	–	–	–	–
γ-muurolene, 1480	–	–	–	–	–	–	–	–	–	0.22	–	–
ar-curcumene, 1481	–	–	–	–	–	–	–	–	–	0.04	–	–
Germacrene D, 1485	–	–	–	–	–	–	–	–	–	0.13	–	–
Valencene, 1496	–	–	–	–	–	–	–	–	–	–	0.21	0.15
α-muurolene, 1500	–	–	–	–	–	–	–	–	–		–	0.22
Cuparene, 1505	–	–	–	–	–	–	–	–	–	0.22	–	–
E, E-α-farnesene, 1506	–	–	–	–	–	–	–	–	–	–	0.23	–
β-bisabolene, 1506	0.88	1.31	1.19	0.56	–	–	–	–	–	–	–	–
Cis-α-bisabolene, 1507	–	–	–	–	–	–	–	–	–	–	–	–
γ-cadinene 1514	–	–	–	–	–	–	–	–	–	0.11	0.25	0.29
δ-cadinene, 1523	–	–	–	0.27	–	–	–	–	–	0.21	0.97	0.77
α-cadinene, 1539	–	–	–	–	–	–	–	–	–	0.05	–	–
α-calacorene, 1546	–	–	–	–	0.1	tr	–	–	–	–	0.29	0.12
Elemol, 1550	0.03	–	0.06	–	–	–	–	–	–	–	–	–
Salviadienol	–	–	–	–	tr	0.7	–	–	–	–	–	–
Cis-nerolidol, 1563	–	–	–	–	–	–	–	–	–	–	–	–
Trans-nerolidol	0.05	–	0.03	–	–	4.2	–	–	–	–	–	–
β-calacorene, 1566	–	–	–	–	0.1	–	–	–	–	–	–	–
Dodecanoic acid, 1567	0.57	1.46	–	–	–	–	–	–	–	–	–	–
Spathulenol, 1578	–	–	–	0.28	9.0	37.6	–	–	–	11.88	22.36	13.26
β-copaene-4-α-ol, 1591	–	–	–	–	0.2	–	–	–	–	–	–	–
Nor-copanone	–	–	–	–	–	–	–	–		1.24	0.52	0.06

Table A.1 (continued)

Plants name	*S. bachtarica* [66]			*S. cuneifolia* [75]	*S. subspicata* [160]		*S. darwinii* [34]	*S. multiflora* [34]	*S. punctata* ssp. *punctata* [102]	*S. biflora* [93]	*S. masukensis* [93]	*S. pseudosimensis* [93]
Compounds name	Leaf	Stem	Aerial part		Summer	Autumn						
Decanal	1.21	2.63	–	–	–	–	–	–	–	–	–	–
p-vinylguaiacol	–	–	–	–	–	–	0.9	–	–	–	–	–
Isospathulenol	–	–	–	–	0.6	0.6	–	–	–	–	2.49	0.23
Spathulenol, 1578	0.63	0.10	–	–	–	–	–	–	–	–	–	–
Caryophylene oxide, 1583	1.58	2.09	1.37	0.38	2.4	6.8	–	–	–	–	–	–
Caryophylla-3(15),7(14)-diene-6-α-ol	–	–	–	–	0.1	1.5	–	–	–	–	–	–
Caryophylla-3(15),7(14)-diene-6-β-ol	–	–	–	–	0.5	3.0	–	–	–	–	–	–
Caryophyllene epoxide	0.43	0.19	0.42	–	–	–	–	–	–	2.26	–	–
14-hydroxy-β-caryophyllene	–	–	–	–	0.3	1.0	–	–	–	–	–	–
14-hydroxy-9-epi-β-caryophyllene	–	–	–	–	0.2	2.7	–	–	–	–	–	–
Viridiflorol 1593	–	–	–	–	–	3.1	–	–	–	–	–	–
Salvial-4(14)-en-1-one, 1595	–	–	–	–	0.6	tr	–	–	1.43	0.90	–	0.23
Torilenol, 1599	–	–	–	–	0.3	1.2	–	–	–	–	–	–
Guaiol, 1601	0.59	–	–	–	–	–	–	–	–	–	–	–
Humulene epoxide II, 1606	–	–	–	–	–	2.0	–	–	–	–	–	–
10-epi-γ-eudesmol, 1624	0.45	0.08	0.03	–	–	–	–	–	–	–	–	–
1-epi-cubenol, 1629	–	–	–	–	0.2	0.7	–	–	–	–	–	–
Epi-α-muurolol, 1642	–	–	–	–	–	–	–	–	–	–	0.40	0.30
3-iso-thujospanone, 1643	–	–	–	–	0.3	–	–	–	–	–	–	–
α-muurolol, 1646	–	–	–	–	–	–	–	–	–	–	–	0.21
Cubenol, 1647	–	–	–	–		1.7	–	–	–	–	–	–

Table A.1 (continued)

Plants name	*S. bachtarica* [66]			*S. cuneifolia* [75]	*S. subspicata* [160]		*S. darwinii* [34]	*S. multiflora* [34]	*S. punctata* ssp. *punctata* [102]	*S. biflora* [93]	*S. masukensis* [93]	*S. pseudosimensis* [93]
Compounds name	Leaf	Stem	Aerial part		Summer	Autumn						
Agarospirol, 1648	0.83	0.94	0.09	–	–	–	–	–	–	–	–	–
β-eudesmol, 1651	0.90	0.92	0.36	–	–	–	–	–	–	–	–	–
Epi-α-cadinol, 1654	–	–	–	–	tr	3.3	–	–	–	–	–	–
α-cadinol 1654	–	–	–	–	0.4	6.1	–	–	–	–	3.54	0.63
α-bisabolol, 1686	1.93	1.05	0.51	–	–	–	–	–	–	1.78	–	–
Eudesma-4(15),7-diene-1-ol, 1688	–	–	–	–	0.1	8.3	–	–	–	–	–	–
α-bisabolol oxide, 1655	–	–	–	–	–	–	–	–	–	8.77	–	–
Junicedranol, 1693	–	–	–	–	0.1		–	–	–	–	–	–
Farnesol, 17,,	0.03	–	–	–	–	–	–	–	–	–	–	–
14-hydroxy-α-humulene, 1714	–	–	–	–	0.3	0.9	–	–	–	–	–	–
oplopanone, 1734	–	–	–	–	0.2		–	–	–	–	–	–
n-tetradecanoic acid, 1720	1.74	3.05	0.09	–	tr	0.2	–	–	–	–	–	–
Aristolone, 1756	–	–	–	–	–	–	–	–	–	0.52	–	–
Benzyl benzoate, 1762	–	–	–	–	0.2	0.5	–	–	–	–	–	–
14-hydroxy-α-muurolene, 1780	–	–	–	–	tr	0.1	–	–	–	–	–	–
Farnesyl acetone, 1918	0.52	0.88	0.05	–	–	–	–	–	–	–	–	–
Isophytol, 1948	–	–	–	–	–	–	–	–	–	0.05	–	–
n-hexadecanoic acid, 1950	5.13	9.49	0.87	–	tr	2.6	–	–	–	–	–	–
n-tetracosane, 2400	–	–	–	–	0.1	0.3	–	–	–	–	–	–
n-pentacosane, 2500	–	–	–	–	0.2	0.5	–	–	–	–	–	–
n-hexacosane, 2600	–	–	–	–	0.2	0.5	–	–	–	–	–	–

tr trace

Table A.1 (continued)

Plants name	*S. cuneifolia* flowering period				*S. cuneifolia* different part of Croatia [58]									S. cuneifolia [69, 72]	
Compounds name	During (Iran) [108]	Before [46]	During [46]	After [46]	Biokovo			Brac			Kozjak			Turkey	
					July	Sept	Nov	July	Sept	Nov	July	Sept	Nov		
α-thujene, 930	–	–	–	–	1.8	1.8	1.2	1.4	1.3	1.3	1.4	1.7	1.5	–	0.37
α-pinene, 939	–	8.1	5.8	12.0	tr	1.0	1.3	0.6	1.4	2.2	0.4	1.3	1.0	–	0.93
Camphene, 954	–	–	–	–	–	–	–	–	–	–	–	–	–	–	0.33
β-pinene, 979	–	1.5	–	3.4	–	–	–	–	–	–	–	–	–	–	0.17
1-octen-3-ol, 979	–	0.7	–	–	0.5	0.7	1.5	0.4	1.0	1.4	0.6	1.3	1.1		0.23
Myrcene, 991	–	2.8	0.3	0.9	0.6	0.8	–	0.9	0.6	–	1.1	1.3	1.7	2.1	1.57
α-phellandrene, 1003	–	–	–	–	–	–	–	–	–	–	–	–	–	–	0.17
α-terpinene, 1017	0.2	–	–	–	–	–	–	–	–	–	–	–	–	2.1	1.83
p-cymene, 1025	1.7	1.8	14.8	4.0	3.0	12.6	28.9	3.8	15.2	25.6	5.0	17.8	19.1	7.3	16.28
Limonene, 1029	–	4.7	1.8	11.0	–	–	1.0	–	–	1.1	–	1.6	1.6	–	0.56
1,8-cineole, 1031	–	–	–	–	–	–	–	–	–	–	–	–	–	–	0.52
Cis-β-ocimene, 1037	–	4.2	–	–	–	–	–	–	–	–	0.4	0.6	0.6	–	2.33
Trans-β-ocimene, 1050	–	–	–	–	–	–	–	–	–	–	–	–	–	–	0.60
γ-terpinene, 1060	1.1	4.1	–	–	7.2	8.1	–	5.8	4.9	–	6.5	7.6		27.4	12.59
Cis-sabinene hydrate, 1070	–	0.5	–	–	–	–	–	–	–	–	–	–	–	–	0.12
Cis-linalool oxide 1087	1.7	–	–	–	–	–	–	–	–	–	–	–	–	–	–
Terpinolene, 1089	–	–	–	–	–	–	–	–	–	–	–	–	–	–	0.26
Trans-sabinene hydrate, 1098	–	0.5	–	–	0.3	–	1.6	0.3	0.2	1.2	0.3	0.4	0.5	–	
Linalool, 1097	22.9	18.2	17.2	17.9	24.8	0.5	0.7	0.5	0.6	0.9	0.8	0.7	0.7	–	0.36

Table A.1 (continued)

Plants name	*S. cuneifolia* flowering period				*S. cuneifolia* different part of Croatia [58]									S. cuneifolia [69, 72]	
Compounds name	During (Iran) [108]	Before [46]	During [46]	After [46]	Biokovo			Brac			Kozjak			Turkey	
					July	Sept	Nov	July	Sept	Nov	July	Sept	Nov		
Allo-ocimene, 1132	–	2.6	–	–	0.9	–	–	0.2			0.5	0.5	0.5	–	–
α-copaene, 1377	–	–	–	0.7	–	–	–	–	–	–	–	–	–	–	–
Trans-linalool oxide	1.6	–	–	–	–	–	–	–	–	–	–	–	–	–	–
Terpinen-4-ol, 1177	0.7	2.8	5.1	6.4	–	–	–	–	–	–	–	–	–	–	1.22
Camphor, 1146	–	1.4	–	1.9	–	–	0.4	–	–	0.7	–	–	0.5	–	–
Geraniol, 1153	–	0.4	1.0	–	tr	–	–	–	10.2	–	9.2	9.7	9.5	–	–
Borneol, 1169	1.1	6.8	7.6	7.4	1.5	4.8	8.1	3.2	6.5	11.5	3.0	5.8	6.2	0.7	0.89
p-cymen-8-ol, 1183	–	–	–	–	–	–	–	–	–	–	–	–	–	–	0.13
Geranial, 1167	–	–	–	–	–	–	–	–	0.9	–	0.7	1.1	1.4	–	–
α-terpineol, 1198	0.1	–	1.9	1.9	–	0.5	0.6	–	0.3	0.6		0.3	0.4	–	0.28
Calaren	–	–	1.1	–	–	–	–	–	–	–	–	–	–	–	–
Nerol, 1235	0.3	–	–	–	–	–	–	–	0.9	0.2	1.5	1.0	1.2	–	–
Thymol methyl ether, 1235	–	–	–	–	6.8	2.3	5.2	3.2	3.3	12.8	1.5	4.1	3.3	–	0.07
Neral, 1238	–	2.0	–	2.8	–	–	–	–	0.6	–	–	0.7	1.1	–	–
Cuminaldehyde, 1242	–	–	–	–	–	–	–	–	–	–	–	–	–	–	–
Carvacrol methyl ether, 1245	0.6	–	–	–	2.0	11.0	7.4	4.6	6.4	5.4	7.0	8.2	4.2	–	–
Bornyl acetate, 1289	–	–	–	–	–	–	–	–	–	–	–	–	–	0.1	tr
Thymol, 1290	0.7	1.8	1.6	0.8	15.0	3.9	3.0	11.0	5.4	2.6	20.6	6.1	1.9	0.1	8.34
Carvacrol, 1299	57.9	5.0	16.3	7.1	25.3	45.7	28.1	52.4	26.2	16.1	30.4	20.8	23.4	53.3	45.39
Thymol acetate, 1352	–	–	–	–	tr	0.2	–	0.2	–	–	0.2	–	–	–	–
Geranyl acetate, 1381	–	–	–	–	–	–	–	–	5.2	–	3.5	2.5	6.7	–	–

Table A.1 (continued)

Plants name	*S. cuneifolia* flowering period				*S. cuneifolia* different part of Croatia [58]									S. cuneifolia [69, 72]	
Compounds name	During (Iran) [108]	Before [46]	During [46]	After [46]	Biokovo			Brac			Kozjak			Turkey	
					July	Sept	Nov	July	Sept	Nov	July	Sept	Nov		
β-bourbonene, 1388	–	–	2.2	3.2	–	–	–	–	–	0.3	–	–	0.6	–	–
β-cubebene, 1388	–	9.1	3.5	1.7	0.4	2.0	–	0.5	–	1.5	1.0	0.5	0.5	–	–
α-gurjunene 1410	0.3	–	–	–	–	–	–	–	–	–	–	–	–	–	–
Aromadendrene, 1441	–	–	–	–	tr	tr	–	0.2	0.3	0.5	0.2	0.3	–	–	–
α-humulene, 1455	0.2	–	–	–	0.2	–	–	0.2	–	–	–	–	–	–	0.09
β-caryophyllene, 1466	2.8	5.2	9.3	2.4	2.7	–	–	1.8	1.3	1.3	–	–	–	–	0.49
Germacrene D, 1485	–	–	–	–	–	–	–	–	–	–	–	–	–	–	0.17
β-selinene, 1490	0.2	–	–	–		–	–	–	–	–	–	–	–	–	–
α-selinene, 1498	0.2	–	–	–	–	–	–	–	–	–	–	–	–	–	–
Cumin alcohol	–	–	–	–	–	–	–	–	–	–	–	–	–	–	0.14
β-bisabolene, 1506	–	–	–	0.6	0.4	2.0		0.5	–	1.5	1.0	0.5	0.5	–	0.97
δ-cadinene, 1523	–	–	–	–	0.1	0.4	0.2	0.3	–	0.2	–	–	–	–	–
Germacrene B, 1561	–	–	–	–	–	–	–	–	–	–	–	–	–	–	tr
Spathulenol, 1578	–	1.3	1.5	1.9	0.3.	0.3	0.5	–	0.3	0.7	0.4	0.2	0.7	–	–
Caryophylene oxide, 1583	2.3	0.4	1.4	1.8	–	0.4	2.6	–	1.2	2.8	0.5	0.6	2.3	–	tr
Viridiflorol 1593	–	0.3	–	–	–	–	–	–	–	–	–	–	–	–	–
β-eudesmol, 1651	0.2	–	–	–	–	–	–	–	–	–	–	–	–	–	–
7-epi-α-eudesmol, 1664	0.2	–	–	–	–	–	–	–	–	–	–	–	–	–	–

Table A.1 (continued)

Plants name	*S. montana*								*S. montana* different part of Croatia [58]								
Compounds name	Croatia [81]	Italy [126, 127, 144, 161, 165]				Croatia [49]			Biokovo			Brac			Kozjak		
						Before flowering		Flowering	July	Sept	Nov	July	Sept	Nov	July	Sept	Nov
						June	Aug	Sept									
α-thujene, 930	–	1.4	–	0.9	–	1.77	1.92	1.58	1.8	1.8	1.2	1.4	1.3	1.3	1.4	1.7	1.5
α-pinene, 939	3.34	0.6	1.07	1.8	0.8	0.52	1.29	0.99	tr	1.0	1.3	0.6	1.4	2.2	0.4	1.3	1.0
Camphene, 954	2.20	–	–	0.7	0.2	–	–	–	–	–	–	–	–	–	–	–	–
β-pinene, 979	2.18	–	–	2.0	0.2	0.1	tr	tr	–	–	–	–	–	–	–	–	–
1-octen-3-ol, 979	–	0.4	–	–	–	0.55	0.87	0.63	0.5	0.7	1.5	0.4	1.0	1.4	0.6	1.3	1.1
Sabinene, 975	–	–	10.03	–	–	–	–	–	–	–	–	–	–	–	–	–	–
Myrcene, 991	2.95	0.9	–	0.8	1.9	2.02	2.60	1.42	0.6	0.8	–	0.9	0.6	–	1.1	1.3	1.7
α-terpinene, 1017	–	1.7	1.41	0.8	2.0	3.87	2.34	1.71	1.8	1.5	–	1.7	0.9	–	1.0	1.8	–
p-cymene, 1025	14.30	4.8	9.83	41.4	9.7	7.10	13.48	12.36	3.0	12.6	28.9	3.8	15.2	25.6	5.0	17.8	19.1
β-phellandrene, 1030	5.60	–	–	tr	–	0.44	0.35	tr	–	–	–	–	–	–	–	–	–
Limonene, 1029	–	–	–	0.7	0.4	1.68	tr	tr	–	–	1.0	–	–	1.1	–	1.6	1.6
δ-3-carene, 1031	0.3	–	–	tr	–	0.4	0.1	tr	–	–	–	–	–	–	–	–	–
1,8-cineole, 1031	–	–	–	0.5	0.3	–	–	–	–	–	–	–	–	–	–	–	–
Cis-β-ocimene, 1037	0.21	–	–	–	–	–	–	–	–	–	–	–	–	–	0.4	0.6	0.6
Trans-β-ocimene, 1050	0.04	–	–	–	–	–	–	–	–	–	–	–	–	–	–	–	–
γ-terpinene, 1060	–	5.8	–	3.0	13.2	9.49	9.74	7.57	7.2	8.1		5.8	4.9	–	6.5	7.6	–
Cis-sabinene hydrate, 1070	–	–	–	tr	–	–	–	–	–	–	–	–	–	–	–	–	–
α-terpinolene, 1089	0.34		–	–	–	tr	0.29	tr	–	–	–	–	–	–	–	–	–
Linalool, 1097	4.81	0.5		–	0.3	0.37	3.05	3.15	24.8	0.5	0.7	0.5	0.6	0.9	0.8	0.7	0.7

Table A.1 (continued)

Plants name	S. montana								S. montana different part of Croatia [58]								
Compounds name	Croatia [81]	Italy [126, 127, 144, 161, 165]				Croatia [49]			Biokovo			Brac			Kozjak		
						Before flowering		Flow-ering	July	Sept	Nov	July	Sept	Nov	July	Sept	Nov
						June	Aug	Sept									
Trans-sabinene hydrate, 1098	–	0.3	–	–	–	–	–	–	0.3		1.6	0.3	0.2	1.2	0.3	0.4	0.5
Thujone 1102 1114	0.11	–	–	–	–	–	–	–	–	–	–	–	–	–	–	–	–
Allo-ocimene, 1132	–	0.9	–	–	–	0.96	–	–	0.9	–	–	0.2	–	–	0.5	0.5	0.5
Camphor, 1146	2.03	–	–	–	–	–	–	–	–	–	0.4	–	–	0.7	–	–	0.5
Geraniol, 1153	1.41	–	–	–	–	–	–	–	–	–	–	–	–	–	–	–	–
β-terpineol, 1163 1144	1.76	–	–	–	–	–	–	–	–	–	–	–	–	–	–	–	–
Borneol, 1169	–	–	2.36	–	0.7	2.36	4.05	3.62	1.5	4.8	8.1	3.2	6.5	11.5	3.0	5.8	6.2
Terpinen-4-ol, 1177	2.66	–	–	2.0	0.1	0.14	0.11	tr	–	–	–	–	–	–	–	–	–
p-cymen-8-ol, 1183	–	–	–	–	–	–	0.11	–	–	–	–	–	–	–	–	–	–
p-cymen-4-ol	–	–	–	0.8	–	–	–	–	–	–	–	–	–	–	–	–	–
Myrtenol, 1196	–	–	–	–	–	–	–	–	–	–	–	–	–	–	–	–	–
Geranial, 1167	–	–	–	–	–	–	6.43	4.15	tr	–	–	–	1.02	–	9.2	9.7	9.5
α-terpineol, 1198	0.13	–	–	2.6	–	–	–	–	–	0.5	0.6	–	0.3	0.6	–	0.3	0.4
Thymol methyl ether, 1235	–	3.2	–	–	–	5.11	6.83	4.99	6.8	2.3	5.2	3.2	3.3	12.8	1.5	4.1	3.3
Pulegone, 1237	2.14	–	–	–	–	–	–	–	–	–	–	–	–	–	–	–	–
Neral, 1238	–	–	–	–	–	–	tr	0.45	–	–	–	–	0.6	–	–	0.7	1.1
Carvone, 1243	–	–	0.41	–	–	–	–	–	2.7	–	–	1.8	1.3	1.3	–	–	–
Carvacrol methyl ether, 1245	–	4.6	–	–	–	4.60	4.01	5.11	2.0	11.0	7.4	4.6	6.4	5.4	7.0	8.2	4.2

Table A.1 (continued)

Plants name	*S. montana*								*S. montana* different part of Croatia [58]								
Compounds name	Croatia [81]	Italy [126, 127, 144, 161, 165]				Croatia [49]			Biokovo			Brac			Kozjak		
						Before flowering		Flowering	July	Sept	Nov	July	Sept	Nov	July	Sept	Nov
						June	Aug	Sept									
Thymoquinone, 1252	–	–	–	0.8	–	–	–	–	–	–	–	–	–	–	–	–	–
Thymol, 1290	–	11.0	0.31	tr	0.3	46.02	30.88	35.41	15.0	3.9	3.0	11.0	5.4	2.6	20.6	6.1	1.9
Carvacrol, 1299	–	50.2	56.82	37.0	56.8	4.52	3.81	6.86	25.3	45.7	28.1	52.4	26.2	16.1	30.4	20.8	23.4
Thymol acetate, 1352	–	–	–	–	–	0.17		–	tr	0.2	–	0.2	–	–	0.2	–	–
Carvacrol acetate, 1373	–	–	–	–	–	0.23	0.1	–		–	–	–	–	–	–	–	–
β-cubebene, 1388	–	–	–		–	0.30	tr	tr	–	0.5	–	0.4	–	0.7	0.4	0.4	0.5
β-bourbonene, 1388	–	–	–	–	–	–	–	–	–	–	–	–	–	0.3	–	–	0.6
Aromadendrene, 1441	–	–	–	–	–	0.38	0.2	0.22	tr	tr	–	0.2	0.3	0.5	0.2	0.3	
α-humulene, 1455	–	–	–	–	–	–	–	–	0.2	–	–	0.2	–	–	–	–	–
β-caryophyllene, 1466	–	–	1.44	1.6	3.6	3.75	1.66	2.64	–	–	–	–	–	–	–	–	–
Germacrene D, 1485	–	–	–	–	0.4	–	–	–	–	–	–	–	–	–	–	–	–
β-bisabolene, 1506	–	–	–	0.5	–	–	–	–	–	–	–	–	–	–	–	–	–
Spathulenol, 1578	–	–	–	–	–	0.34	–	–	0.3	0.3	0.5	–	0.3	0.7	0.4	0.2	0.7
Caryophylene oxide, 1583	–	–	–	1.0	–	–	–	–	–	0.4	2.6	–	1.2	2.8	0.5	0.6	2.3
Ledene (viridiflorene 1593)	–	–	–	–	–	1.28	0.29	0.34	–	–	–	–	–	–	–	–	–

tr trace

Table A.1 (continued)

Plants name	*S. obovata* [35]								*S. brevicalix* [2]
Compounds name	1989			1990			1991		
	Flowering	Fruiting	Post fruiting	Flowering	Fruiting	Post fruiting	Flowering	Post fruiting	
α-thujene, 930	tr	0.5	tr	tr	0.79	tr	tr	tr	0.1
Artemisiatriene, 929	tr	–	–	–	–	–	0.75	–	–
α-pinene, 939	2.52	0.84	tr	3.8	3.01	2.05	2.11	1.45	0.1
1-tetradecyne	–	–	–	–	–	–	–	–	0.1
Camphene, 954	5.21	1.66	1.68	7.83	5.57	3.12	4.78	2.31	–
Sabinene, 975	tr	tr	3.58	tr	0.6	tr	0.55	tr	0.3
1-octen-3-ol, 979	–	–	–	–	–	–	–	–	tr
β-pinene, 979	0.57	tr	tr	0.98	0.88	tr	0.61	tr	–
Myrcene, 991	1.81	0.71	tr	1.52	0.76	2.61	–	–	0.4
α-phellandrene, 1003	0.54	tr	tr	tr	tr	2.14	tr	tr	–
α-terpinene, 1017	–	–	–	–	–	–	–	–	–
p-cymene, 1025	3.52	4.31	5.64	9.61	8.24	6.32	7.06	4.39	2.3
Limonene, 1029	–	–	–	–	–	–	–	–	–
β-phellandrene, 1030	tr	tr	tr	0.60	0.80	1.10	1.14	1.60	–
δ-3-carene, 1031	tr	tr	tr	tr	tr	tr	–	–	–
1,8-cineole, 1031	–	–	–	–	–	–	–	–	2.8
Cis-β-ocimene, 1037	–	–	–	–	–	–	–	–	0.1
Trans-β-ocimene, 1050	–	–	–	–	–	–	–	–	0.1
γ-terpinene, 1060	1.92	1.13	tr	2.28	2.41	9.22	3.61	7.60	0.1
Cis-sabinene hydrate, 1070	0.61	tr	tr	0.62	0.70	tr	tr	tr	
Cis-linalool oxide 1087	0.60	tr	tr	0.63	2.64	ttr	tr	tr	tr
Terpinolene, 1089	tr	tr	tr	0.59	tr	tr	1.09	1.03	0.1
Myrtenol, 1196	–	–	–	–	–	–	–	–	–
Linalool, 1097	4.43	1.38	0.75	1.77	tr	10.65	7.82	8.59	0.6
Octane-3-yl acetate, 1113	–	–	–	–	–	–	–	–	0.4

Table A.1 (continued)

Plants name	*S. obovata* [35]								*S. brevicalix* [2]
Compounds name	1989			1990			1991		
	Flowering	Fruiting	Post fruiting	Flowering	Fruiting	Post fruiting	Flowering	Post fruiting	
Cis-*p*-menth-2-ene-1-ol, 1122	–	–	–	–	–	–	–	–	0.1
Allo-ocimene, 1132	–	–	–	–	–	–	–	–	–
Trans-*p*-menth-2-ene-1-ol, 1141	–	–	–	–	–	–	–	–	tr
β-terpineol, 1163 1144	1.22	tr	tr	0.75	1.61	0.39	1.39	1.39	–
Camphor, 1146	14.85	23.30	39.93	31.48	27.72	9.59	18.22	11.78	–
Menthone, 1153	–	–	–	–	–	–	–	–	37.5
Geraniol, 1153	tr	tr	tr	tr	tr	tr	tr	tr	–
Terpinen-4-ol, 1177	2.32	3.44	3.15	1.79	2.21	1.58	2.29	3.66	–
p-cymen-8-ol, 1183	0.85	0.68	1.06	tr	1.04	tr	tr	tr	–
Cis-dihydrocarvone, 1193	–	–	–	–	–	–	–	–	5.0
Cis-piperitol, 1196	1.68	1.66	1.24	1.05	1.33	1.53	1.87	2.08	–
α-terpineol, 1198	11.90	4.50	4.27	1.69	2.21	9.56	7.89	9.45	tr
Trans-carveol, 1217	–	–	–	–	–	–	–	–	0.4
Cis-cinnamaldehyde, 1219	–	–	–	–	–	–	–	–	0.4
Citronellol, 1228	tr	tr	tr	tr	tr	tr	tr	tr	–
Pulegone, 1237	–	–	–	–	–	–	–	–	8.4
Cis-citral, 1240	tr	tr	tr	tr	tr	tr	tr	tr	–
Piperitone, 1253	–	–	–	–	–	–	–	–	2.4
Isomenthone	–	–	–	–	–	–	–	–	25.2
Linalyl acetate 1257	tr	tr	0.60	tr	0.58	tr	tr	–	0.4
Perilla aldehyde, 1272	–	–	–	–	–	–	–	–	0.3
Bornyl acetate, 1289	tr	tr	0.83	0.74	0.60	tr	1.30	–	–
γ-caryophyllne-isocaryophyllene	2.53	1.87	2.03	2.02	3.06	1.50	2.77	2.30	–
Thymol, 1290	5.65	10.30	5.73	5.29	5.04	10.38	11.20	12.87	2.3
Iso-dihydrocarveol, 1293	–	–	–	–	–	–	–	–	tr

Table A.1 (continued)

Plants name	*S. obovata* [35]								*S. brevicalix* [2]
Compounds name	1989			1990			1991		
	Flowering	Fruiting	Post fruiting	Flowering	Fruiting	Post fruiting	Flowering	Post fruiting	
Carvacrol, 1299	1.29	1.57	1.80	0.95	1.45	0.99	0.98	1.14	0.2
α-cubebene, 1351	tr	0.53	0.54	0.97	0.94	1.24	2.28	3.48	tr
Thymol acetate, 1352	–	–	–	–	–	–	–	–	0.4
β-cubebene, 1388	–	–	–	–	–	–	–	–	tr
β-bourbonene, 1388	–	–	–	–	–	–	–	–	0.1
β-elemene, 1391	–	–	–	–	–	–	–	–	tr
α-humulene, 1455	–	–	–	–	–	–	–	–	0.1
Alloaromadendrene, 1461	–	–	–	–	–	–	–	–	0.1
β-caryophyllene, 1466	0.75	3.19	11.12	7.14	7.87	0.73	0.56	tr	1.0
γ-muurolene, 1480	–	–	–	–	–	–	–	–	0.1
α-bisabolene, 1507	tr	tr	tr	tr	0.50	tr	tr	tr	–
Cumin alcohol	–	–	–	–	–	–	–	–	1.0
Ledol, 1569	tr	tr	tr	tr	tr	tr	tr	tr	–
Germacrene B, 1561	1.73	0.99	0.59	tr	tr	0.88	1.76	2.32	–
Caryophylene oxide, 1583	1.12	1.51	1.73	0.51	1.33	tr	1.64	0.77	–
Viridiflorol 1593	–	–	–	–	–	–	–	–	tr
Guaiol, 1601	–	–	–	–	–	–	–	–	0.1
γ-eudesmol, 1632	–	–	–	–	–	–	–	–	0.1
Torreyol, 1645	–	–	–	–	–	–	–	–	tr
β-eudesmol, 1651	–	–	–	–	–	–	–	–	tr
α-eudesmol 1654	–	–	–	–	–	–	–	–	0.1
α-cadinol 1654	–	–	–	–	–	–	–	–	0.1
Oplopanone, 1734	–	–	–	–	–	–	–	–	0.4
Sclareol, 2223	–	–	–	–	–	–	–	–	0.4

tr trace

Table A.1 (continued)

Plants name	*S. hortensis*											
Compounds name	Iran, [44]				Iran [6]	Turkey [82]	Czech [151]	UK [36]			Syria [43]	
	FC	LS1	LS2	HS				July	Aug	Sept	Old leaf	Young leaf
α-thujene, 930	1.4	1.7	1.6	1.9	1.9	0.8	0.9	–	–	–	–	–
α-pinene, 939	1.3	1.6	1.5	2.1	1.3	2.6	0.9	0.4	0.3	0.4	–	–
Camphene, 954	–	–	–	–	–	–	–	2.2	0.8	0.9	–	–
Sabinene, 975	0.2	0.2	0.2	0.3	–	–	–	–	–	–	–	–
β-pinene, 979	0.7	0.9	1.0	1.4	0.7	2.7	0.4	–	–	–	–	–
Myrcene, 991	–	–	–	–	2.4	1.7	1.4	2.4	3.9	3.0	2.3	2.2
α-phellandrene, 1003	0.3	0.4	0.3	0.4	0.5	0.2	0.2	–	–	–	–	–
α-terpinene, 1017	4.9	4.9	5.1	5.4	4.3	2.2	3.5	0.9	1.2	1.1	3.2	3.4
p-cymene, 1025	2.2	2.2	2.2	2.7	–	9.3	4.7	–	–	–	1.9	1.9
Limonene, 1029	0.6	1.1	0.6	0.7	–	–	–	–	–	–	–	–
δ-3-carene, 1031	–	–	–	–	–	–	0.5	1.5	2.2	1.8	–	–
β-ocimene	–	–	–	–	–	0.1	–	4.9	2.8	5.8	–	–
β-phellandrene, 1030	–	–	–	–	–	–	0.1	–	–	–	–	–
1,8-cineole, 1031	–	–	–	–	–	–		–	–	–	–	–
Trans-β-ocimene, 1050	0.1	0.1	–	0.1	–	–	–	–	–	–	–	–
γ-terpinene, 1060	40.9	39.5	38.3	37.8	31.8	22.6	36.7	35.5	44.6	37.8	3.7	31.5
α-terpinolene, 1089	–	–	–	–	–	0.1	–	–	–	–	–	–
Linalool, 1097	–	–	–	–	–	–	–	0.6	0.7	0.6	–	–
Menthone, 1153	–	–	–	–	–	–	–	–	–	–	–	–
Endo-borneol, 1160	–	–	–	–	–	–	–	–	–	–	–	–

Table A.1 (continued)

Plants name	*S. hortensis*											
Compounds name	Iran, [44]				Iran [6]	Turkey [82]	Czech [151]	UK [36]			Syria [43]	
	FC	LS1	LS2	HS				July	Aug	Sept	Old leaf	Young leaf
Borneol, 1169	–	–	–	–	–	0.3		–	–	–	–	–
Terpinen-4-ol, 1177	–	–	–	–	–	0.3	0.1	–	–	–	–	–
Myrtenol, 1196	–	–	–	–	–	–	–	0.4	0.5	0.5	–	–
Pulegone, 1237	–	–	–	–	–	0.2	–	–	–	–	–	–
Carvacrol methyl ether, 1245	–	–	–	–	0.4	–	–	–	–	–	–	–
Anethole, 1285 1253	–	–	–	–	–	0.1	–	–	–	–	–	–
Piperitone, 1253	–	–	–	0.9	–	–	–	–	–	–	–	–
Thymol, 1290	–	–	–	–	–	29.0	–	–	–	–	–	–
Carvacrol, 1299	41.3	42.6	44.5	40.3	33.7	26.5	48.1	48.5	40.6	44.9	58.9	59.1
Thymol acetate, 1352	1.7	0.8	0.7	1.2	–	0.3	–	–	–	–	–	–
Carvacrol acetate, 1373	–	–	–	–	–	0.1	–	–	–	–	–	–
Trans-caryophyllene, 1419	–	–	–	–	–	–	0.7	–	–	–	–	–
Aromadendrene, 1441	–	–	–	–	–	0.1	–	–	–	–	–	–
β-caryophyllene, 1466	1.5	1.3	1.3	1.2	–	–	–	–	–	–	–	–
Bicyclogermacrene, 1500	–	–	–	–	–	0.1	–	–	–	–	–	–
β-bisabolene, 1506	0.9	0.5	0.4	0.5	–	0.2	1.7	–	–	–	2.2	2.3

tr trace, *FC* Irrigation treatment to full field capacity during the growth season, *LS1* Low water stress treatment from vegetative to full flowering stage, *LS2* Low water stress treatment from near to full flowering stage, *HS* Severe water stress treatment from near to full flowering stage

Table A.1 (continued)

Plants name	*S. doulgassi* [37]																			
Compounds name	Carvone type					Isomenthone type					Menthane type					Bicyclic type				
	HT HL	HT LL	LT HL	LT LL	Field	HT HL	HT LL	LT HL	LT LL	Field	HT HL	HT LL	LT HL	LT LL	Field	HT HL	HT LL	LT HL	LT LL	Field
α-pinene, 939	6.7	9.6	4.8	4.8	3.5	11.5	12.6	6.8	5.6	–	12.0	11.0	6.8	6.9	6.9	12.7	11.6	8.5	9.4	9.0
Camphene, 954	12.0	13.9	9.1	9.0	11.1	20.1	18.3	18.3	13.3	15.1	25.1	22.3	22.3	21.2	20.1	24.3	24.6	23.4	26.5	27.6
β-pinene, 979	4.2	5.6	3.0	3.2	2.0	8.3	7.2	3.4	3.3	2.2	6.8	6.8	3.1	3.8	3.2	7.1	6.9	4.5	4.7	3.1
Limonene, 1029	17.2	17.8	22.9	16.3	10.8	5.3	4.5	5.0	3.4	4.4	3.8	3.9	5.8	5.8	6.6	6.7	6.9	10.5	9.8	7.9
1,8-cineole, 1031	0.8	0.3	1.2	0.6	0.1	2.3	0.5	1.6	0.8	1.1	1.3	1.3	1.7	1.3	0.9	1.4	1.3	2.6	2.4	0.6
Menthone, 1153	0.9	1.7	1.0	1.4	0.1	0.2	0.1	0.8	1.9	0.2	2.4	5.4	11.2	17.2	11.9	0.3	0.2	0.5	0.2	0.1
Isomenthone	0.1	–	0.3	1.3	0.2	2.4	7.0	15.8	21.2	31.3	0.9	0.8	0.5	1.2	2.1	–	0.1	0.1	0.2	0.1
Camphor, 1146	21.8	20.7	13.5	13.5	25.2	28.3	25.8	32.5	24.6	20.6	23.0	35.4	35.5	31.2	38.0	34.9	39.3	41.7	39.6	37.4
Terpinene-4-ol	0.5	0.4	0.1	0.2	0.3	0.3	0.2	0.3	0.3	0.1	0.4	0.8	0.4	0.3	0.7	0.9	1.1	1.0	0.8	0.8
Pulegone, 1237	–	–	0.1	0.6	0.1	2.9	4.3	6.0	12.2	9.0	3.6	3.1	2.7	4.5	0.2	–	–	–	–	–
α-terpineole + borneol	2.2	0.6	0.5	0.1	0.6	1.7	0.7	0.7	0.2	0.1	17.6	6.2	6.5	1.6	2.8	8.1	5.7	2.9	3.1	2.6
Carvone, 1243	31.7	28.7	37.7	45.9	42.3	–	–	–	–	–	–	–	–	–	–	–	–	–	–	–
Piperitone, 1253	–	–	–	–	–	6.2	10.8	59	8.9	5.1	0.8	0.8	1.1	0.7	1.7	–	–	0.2	–	0.1
Piperitenone, 1343	–	–	–	–	–	8.1	7.4	0.7	3.3	3.5	–	–	–	–	–	–	–	–	–	–
Other	2.1	1.3	5.2	1.1	3.9	2.7	0.8	2.3	1.0	6.3	2.3	2.3	2.5	2.1	4.9	3.6	4.0	5.6	3.5	10.7

HT high temperature, *HL* high light irradiance, *LT* low temperature, *LL* low light irradiance (average of developmental stages), *field* mature leaves only

References

1. Momtaz S, Abdollahi M. An update on pharmacology of *Satureja* species; from antioxidant, antimicrobial, antidiabetes and antihyperlipidemic to reproductive stimulation. Int J Pharmacol. 2010;6(4):454–61.
2. Jamzad Z. Flora of Iran. Tehran(In Persian): Research Institute of Forests and Rangelands; 2012.
3. Maroofi H. Two new plant species from Kurdistan province, west of Iran. Iran J Bot. 2010;16(1):76–81.
4. Cantino PD, Harley RM, Wagstaff SJ. Generae of Labiatae status and classification. In: Harley RM, Reynolds T, editors. Advances in Labiatae science. Kew: Royal Botanic Garden Press; 1992.
5. Senatore F, Urrunaga-Soria E, Urrunaga-Soria R, Della Porta G, De Feo V Essential oils from two Peruvian *Satureja* species. Flav Fragr J. 1998;13:1–4. http://dx.doi.org/10.1002/(SICI)1099-1026(199801/02)13:1<1::AID-FFJ672>3.0.CO;2-4.
6. American Herbal Pharmacopoeia. Boca Raton: Taylor & Francis; 2011.
7. Deans SG, Svoboda KP. Antibacterial activity of summer savory (*Satureja hortensis* L.) essential oil and its constituents. J Horti Sci. 1989;64:205–10.
8. Zargari A. Medicinal plants. 4th ed. Tehran: Tehran University Publications; 1990.
9. Hajhashemi V, Sadraei H, Ghannadi A, Mohseni M. Antispasmodic and anti-diarrhoel effect of *Satureja hortensis* L. essential oil. J Ethnopharmacol. 2000;71:187–92. http://dx.doi.org/10.1016/S0378-8741(99)00209-3.
10. Macia MJ, Garcia E, Vidaurre PJ. An ethnobotanical survey of medicinal plants commercialized in the markets of La Paz and El Alto, Bolivia. J Ethnopharmacol. 2005;97:337–50. http://dx.doi.org/10.1016/j.jep.2004.11.022.
11. Hilgert NI. Plants used in home medicine in the Zenta River basin, Northwest Argentina. J Ethnopharmacol. 2001;76:11–34. http://dx.doi.org/10.1016/S0378-8741(01)00190-8.
12. Chorianopoulos N, Evergetis E, Mallouchos A, Kalpoutzakis E, Nychas GJ, Haroutounian SA. Characterization of the essential oil volatiles of *Satureja thymbra* and *Satureja parnassica*: influence of harvesting time and antimicrobial activity. J Agric Food Chem. 2006;54:3139–45. http://dx.doi.org/10.1021/jf053183n.
13. Hadji Sharifi A. Mystery of medicinal plants. Tehran: Hafez Novin; 2003.
14. Mosaddegh M. Principles of Iranian traditional medicine. http://www.itmrc.org/principles.htm.
15. Hadji Sharifi A. Mystery of medicinal herbs. 6th ed. Tehran: Lohe Danesh Publisher; 2009.
16. Gabor J, Mathea I, Miklossy-Varia V, Blundenc G. Comparative studies of the rosmarinic and caffeic acid contents of Lamiaceae species. Biochem Syst Ecol. 1999;27:733–8. http://dx.doi.org/10.1016/S0305-1978(99)00007-1.

S. Saeidnia et al., *Satureja: Ethnomedicine, Phytochemical Diversity and Pharmacological Activities,* SpringerBriefs in Pharmacology and Toxicology,
DOI 10.1007/978-3-319-25026-7

17. Hadian J, Ebrahimi SN, Salehi P. Variability of morphological and phytochemical characteristics among *Satureja hortensis* L. accessions of Iran. Ind Crop Prod. 2010;32:62–9. http://dx.doi.org/10.1016/j.indcrop.2010.03.006.
18. Tepe B, Sokmen A. Production and optimization of rosmarinic acid by *Satureja hortensis* L. callus cultures. Nat Prod Res. 2007;21:1133–44. http://dx.doi.org/10.1080/14786410601130737.
19. Zgorka G, Glowniak K. Variation of free phenolic acids in medicinal plants belonging to the Lamiaceae family. J Pharm Biomed Anal 2001;26. http://dx.doi.org/10.1016/s0731-7085(01)00354-5.
20. Exarchou V, Nenadis N, Troganis A, Tsimidou M, Boskou D, Gerothanassis IP. Antioxidant activities and phenolic composition of extracts from Greek oregano, Greek sage, and summer savory. J Agric Food Chem. 2002;50:5294–9. http://dx.doi.org/10.1021/jf020408a.
21. Shekarchi M, Hajimehdipoor H, Saeidnia S, Gohari AR, Pirali Hamedani M. Comparative study of rosmarinic acid content in some plants of Labiatae family. Pharmacogn Mag. 2012;8(29):37–41. http://dx.doi.org/10.4103/0973-1296.93316.
22. Giao SM, Pereira IC, Fonseca CS, Pintado EM, Malcata FX. Effect of particle size upon the extent of extraction of antioxidant power from the plants *Agrimonia eupatoria*, *Salvia* sp. and *Satureja montana*. Food Chem. 2009;117:412–6. http://dx.doi.org/10.1016/j.foodchem.2009.04.020.
23. Kosar M, Dorman HJD, Hiltunen R. Effect of an acid treatment on the phytochemical and antioxidant characteristics of extracts from selected Lamiaceae species. Food Chem. 2005;91:525–33. http://dx.doi.org/10.1016/j.foodchem.2004.06.029.
24. Chkhikvishvili I, Enukidze M, Sanikidze T, Machavariani M, Gogia N, Vinokur Y, et al. Rosmarinic acid-rich extracts of summer savory (*Satureja hortensis* L.) protect jurkat t cells against oxidative stress. Oxid Med Cell Longev. 2013;2013(4):456253. http://dx.doi.org/10.1155/2013/456253.
25. Darbour N, Baltassat F, Raynaud J, Pellecver J. Flavonoid glycosides in leaves of *Satureja hortensis* L. (Labiatae). Pharm Acta Helv. 1990;65:239–40.
26. Leung AY, Foster S. Encyclopedia of common natural ingredients used in food, drugs and cosmetics. 2nd ed. New York: Wiley; 1996.
27. Tomas-Barberan FA, Husain ZS, Gill IM. The distribution of methylated flavones in the Lamiaceae. Biochem Syst Ecol. 1988;16(1):43–6. http://dx.doi.org/10.1016/0305-1978(88)90115-9.
28. Marin DP, Grayer JR, Veitch NC, Kite GC, Harborne JB. Acacetin glycosides as taxonomic markers in Calamintha and Micromeria. Phytochemistry. 2001;58:943–7. http://dx.doi.org/10.1016/S0031-9422(01)00352-1.
29. Skoulaa M, Grayerc JR, Kitec CG. Surface flavonoids in *Satureja thymbra* and *Satureja spinosa* (Lamiaceae). Biochem Syst Ecol. 2005;33:541–4. http://dx.doi.org/10.1016/j.bse.2004.10.003.
30. Escudero J, Lopez JC, Rabanal RM, Valverde S. Secondary methabolites from *Satureja* species. New triterpenoid from *Stureja acinos*. J Nat Prod. 1985;48(1):128–31. http://dx.doi.org/10.1021/np50037a025.
31. Gohari AR, Saeidnia S, Gohari MR, Moradi-Afrapoli F, Malmir M, Hadjiakhoondi A. Bioactive flavonoids from *Satureja atropatana* Bonge. Nat Prod Res. 2009;23(17):1609–14. http://dx.doi.org/10.1080/14786410902800707.
32. Jager S, Trojan H, Kopp T, Laszczyk NM, Scheffler A. Pentacyclic triterpene distribution in various plants-rich sources for a new group of multi-potent plant extracts. Molecules. 2009;14:2016–31. http://dx.doi.org/10.3390/molecules14062016.
33. Rasborsek MI, Voncina DB, Dolecek V, Voncina E. Determination of oleanolic, betulinic and ursolic acid in Lamiaceae and mass spectral fragmentation of their trimethylsilylated derivatives. Chromatographia. 2008;67:433–40. http://dx.doi.org/10.1365/s10337-008-0533-6.
34. Janicsak G, Veres K, Kakasy AZ, Mathe I. Study of the oleanolic and ursolic acid contents of some species of the Lamiaceae. Biochem Syst Ecol. 2006;34:392–6. http://dx.doi.org/10.1016/j.bse.2005.12.004.

35. Gohari AR, Saeidnia S, Hadjiakhoondi A, Abdoullahi M, Nezafati M. Isolation and quantificative analysis of oleanolic acid from *Satureja mutica* Fisch. & C. A. Mey. J Med Plants. 2009;8(5):65–9.
36. Labbe C, Castillo M, Connolly DJ. Mono and sesquiterpenoids from *Satureja gilliesii*. Phytochemistry. 1993;34(2):441–4. http://dx.doi.org/10.1016/0031-9422(93)80026-O.
37. Bianco A, Lamesi S, Passacantilli P. Iridoid glucosides from *Satureja vulgaris*. Phytochemistry. 1984;23(1):121–3. http://dx.doi.org/10.1016/0031-9422(84)83089-7.
38. Niemeyer HM. Composition of essential oils from *Satureja darwinii* (Benth.) Briq. and *S. multiflora* (R. et P.) Briq. (Lamiaceae). Relationship between chemotype and oil yield in *Satureja* spp. J Essent Oil Res. 2010;22(6):477–82. http://dx.doi.org/10.1080/10412905.2010.9700376.
39. Arrebola ML, Navarro MC, Jimenez J, Ocana FA. Variations in yield and composition of the essential oil of *Satureja obovata*. Phytochemistry. 1994;35(1):83–93. http://dx.doi.org/10.1016/S0031-9422(00)90514-4.
40. Svoboda KP, Hay KMR, Waterman PG. Growing summer savory (*Satureja hortensis*) in Scotland: quantitative and qualitative analysis of the volatile oil and factors influencing oil production. J Sci Food Agric. 1990;53:193–202. http://dx.doi.org/10.1002/jsfa.2740530207.
41. Lincoln ED, Langenheim HJ. Effect of light and temperature on monoterpenoid yield and composition in *Satureja douglasii*. Biochem Syst Ecol. 1978;6:21–32. http://dx.doi.org/10.1016/0305-1978(78)90021-2.
42. Rhoades GD, Lincoln ED, Langenheim HJ. Preliminary studies of monoterpenoid variability in *Satureja douglasii*. Biochem Syst Ecol. 1976;4:5–12. http://dx.doi.org/10.1016/0305-1978(76)90003-X.
43. Dardiotia A, Hanlidou E, Lanarasb T, Kokkinia S. The essential oils of the greek endemic *Satureja horvatii* ssp. macrophylla in relation to bioclimate. Chem Biodivers. 2010;7:1968–77. http://dx.doi.org/10.1002/cbdv.200900181.
44. Sefidkon F, Abbasi K, Khaniki GB. Infuence of drying and extraction methods on yield and chemical composition of the essential oil of *Satureja hortensis*. Food Chem. 2006;99:19–23. http://dx.doi.org/10.1016/j.foodchem.2005.07.026.
45. Grosso C, Figueiredo AC, Burillo J, Mainar AM, Urieta JS, Barroso JG, et al. Enrichment of the thymoquinone content in volatile oil from *Satureja montana* using supercritical fluid extraction. J Sep Sci. 2009;32:328–34. http://dx.doi.org/10.1002/jssc.200800490.
46. Kubatova A, Lagadec JMA, Miller JD, Hawthorne BS. Selective extraction of oxygenates from savory and peppermint using subcritical water. Flav Fragr J. 2001;16:64–73. http://dx.doi.org/10.1002/1099-1026(200101/02)16:1<64::AID-FFJ949>3.3.CO;2-4.
47. Novak J, Bahoo L, Mitteregger U, Franz C. Composition of individual essential oil glands of savory (*Satureja hortensis* L., Lamiaceae) from Syria. Flav Fragr J. 2006;21:731–4. http://dx.doi.org/10.1002/ffj.1725.
48. Baher FZ, Mirza M, Ghorbanli M, Rezaii MB. The influence of water stress on plant height, herbal and essential oil yield and composition in *Satureja hortensis* L. Flav Fragr J. 2002;17:275–7. http://dx.doi.org/10.1002/ffj.1097.
49. Sefidkon F, Abbasi K, Jamzad Z, Ahmadi S. The effect of distillation methods and stage of plant growth on the essential oil content and composition of *Satureja rechingeri* Jamzad. Food Chem. 2007;100:1054–8. http://dx.doi.org/10.1016/j.foodchem.2005.11.016.
50. Mirjana S, Nada B, Valerija D. Variability of *Satureja cuneifolia* ten. Essential oils and their antimicrobial activity depending on the stage of development. Eur Food Res Technol. 2004;218:367–71. http://dx.doi.org/10.1007/s00217-003-0871-4.
51. Muller-Riebau JF, Berger MB, Yegen O, Cakir C. Seasonal variations in the chemical compositions of essential oils of selected aromatic plants growing wild in Turkey. J Agric Food Chem. 1997;45:4821–5. http://dx.doi.org/10.1021/jf970110y.
52. Gershenzon J, Lincoln ED, Langenheim HJ. The effect of moisture stress on monoterpenoid yield and composition in *Satureja douglasii*. Biochem Syst Ecol. 1978;6:33–43. http://dx.doi.org/10.1016/0305-1978(78)90022-4.

53. Mastelic J, Jerkovic I. Gas chromatography–mass spectrometry analysis of free and glycoconjugated aroma compounds of seasonally collected *Satureja montana* L. Food Chem. 2002;80:135–40. http://dx.doi.org/10.1016/S0308-8146(02)00346-1.
54. Karousou R, Koureas ND, Kokkini S. Essential oil composition is related to the natural habitats: *coridothymus capitatus* and *Satureja thymbra* in NATURA 2000 sites of Crete. Phytochemistry. 2005;66:2668–73.
55. Baser KHC, Ozek T, Kirimer N, Tümen G. A comparative study of the essential oils of wild and cultivated *Satureja hortensis* L. J Essent Oil Res. 2004;16:422–4. http://dx.doi.org/10.1080/10412905.2004.9698761.
56. Farsam H, Amanlou M, Radpour MR, Salehinia AN, Shafiee A. Composition of the essential oils of wild and cultivated *Satureja khuzistanica* Jamzad from Iran. Flav Fragr J. 2004;19:308–10. http://dx.doi.org/10.1002/ffj.1300.
57. Sefidkon F, Jamzad Z. Chemical composition of the essential of three Iranian *Satureja* species (*S. mutica*, *S. macrantha* and *S. intermedia*). Food Chem. 2005;91:1–4. http://dx.doi.org/10.1016/j.foodchem.2004.01.027.
58. Chorianopoulos N, Kalpoutzakis E, Aligiannis N, Mitaku S, Nychas GJ, Haroutounian SA. Essential oils of *Satureja*, and *Thymus* species: chemical composition and antibacterial activities against foodborne pathogens. J Agric Food Chem. 2004;52:8261–7. http://dx.doi.org/10.1021/jf049113i.
59. Tümen G. The volatile constituents of *Satureja cuneifolia*. J Essent Oil Res. 1991;3:365–6. http://dx.doi.org/10.1080/10412905.1991.9697960.
60. Skocibusic M, Bezic N, Dunkic V. Phytochemical composition and antimicrobial activities of the essential oils from *Satureja subspicata* Vis. Growing in Croatia. Food Chem. 2006;96:20–8. http://dx.doi.org/10.1016/j.foodchem.2005.01.051.
61. Tümen G, Sezik E, Baser KHC. The essential oil of *Satureja parnassica* Heldr. & Sart. ex Boiss subsp. Sipylea P.H. Davies. Flav Fragr J. 1992;7:43–6. http://dx.doi.org/10.1002/ffj.2730070110.
62. Milos M, Radonic A, Bezic N, Dunkic V. Localities and seasonal variations in the chemical composition of essential oils of *Satureja montana* L. and *S. cuneifolia* Ten. Flav Fragr J. 2001;16:157–60. http://dx.doi.org/10.1002/ffj.965.
63. Tümen G, Kirimer N, Ermin N, Baser KHC. The essential oils of two new *Satureja* species from Turkey: *Satureja pilosa* and *S. icarica*. J Essent Oil Res. 1998;10:524–6. http://dx.doi.org/10.1080/10412905.1998.9700959.
64. Muschietti L, van Baren C, Coussio J, Vila R, Clos M, Cañigueral S, Adzet T. Chemical composition of the leaf oil of *Satureja odora* and *Satureja parvofolia*. J Essent Oil Res. 1996;9:681–4. http://dx.doi.org/10.1080/10412905.1996.9701042.
65. Sefidkon F, Abbasi K, Bakhshi-Khaniki G. Influence of drying and extraction methods on yield and chemical composition of the essential oil of *Satureja hortensis*. Food Chem. 2006;99:19–23. http://dx.doi.org/10.1016/j.foodchem.2005.07.026.
66. Sajjadi SE, Baluchi M. Chemical composition of the essential oil of *Satureja boissier*i Hausskn. ex Boiss. J Essent Oil Res. 2002;14:49–50. http://dx.doi.org/10.1080/10412905.2002.9699760.
67. Gasic MJ, Palic R. Monoterpenoids in *Satureja horvatii* Siilic and *Satureja subspicata* Bartl. ex Vis. subsp. subspicata. Bull Soc Chim Beograd. 1983;48:677–9.
68. Economou G, Panagopoulos G, Tarantilis P, Kalivas D, Kotoulas V, Travlos IS, Polysiou M, Karamanos A. Variability in essential oil content and composition of *Origanum hirtum* L., *Origanum onites* L., *Coridothymus capitatus* (L.) and *Satureja thymbra* L. populations from the Greek island Ikaria. Ind Crop Prod. 2011;33:236–41. http://dx.doi.org/10.1016/j.indcrop.2010.10.021.
69. Kurkcuoglu M, Tumen G, Baser CHC. Essential oil constituents of *Satureja boissieri*. Chem Nat Compd. 2001;37(4):329–31. http://dx.doi.org/10.1023/A:1013714316862.
70. Meshkatalsadat MH, Rabiei K, Shabaninejad Y. Chemical characterization of volatile oils of different parts of *Satureja Bachtiarica* Bunge. J Adv Chem. 2013;5(2):678–84.

71. Baydar H, Sagdic O, Ozkan G, Karadogan D. Antibacterial activity and composition of essential oils from *Origanum*, *Thymbra* and *Satureja* species with commercial importance in Turkey. Food Control. 2004;15:169–72. http://dx.doi.org/10.1016/S0956-7135(03)00028-8.
72. Eminagaoglu O, Tepe B, Yumrutas O, Akpula H, Daferera D, Polissiou M, Sokmen A. The *in vitro* antioxidative properties of the essential oils and methanol extracts of *Satureja spicigera* (K. Koch.) Boiss. and *Satureja cuneifolia* ten. Food Chem. 2007;100:339–43. http://dx.doi.org/10.1016/j.foodchem.2005.09.054.
73. Melpomeni Skoula M, Grayer JR. Volatile oils of *Coridothymus capitatus*, *Satureja thymbra*, *Satureja spinosa* and *Thymbra calostachya* (Lamiaceae) from Crete. Flav Fragr J. 2005;20:573–6. http://dx.doi.org/10.1002/ffj.1489.
74. Dunkic V, Bezic N, Vuko E, Cukrov D. Antiphytoviral activity of *Satureja montana* L. ssp. variegata (Host) P. W. Ball essential oil and phenol compounds on CMV and TMV. Molecules. 2010;15:6713–21.
75. Glamoclijaa J, Sokovica M, Vukojevicb J, Milenkovicb I, Van Griensvenc L. Chemical composition and antifungal activities of essential oils of *Satureja thymbra* L. and *Salvia pomifera* ssp. calycina (Sm.) Hayek. J Essent Oil Res. 2006;18(1):115–7.
76. Biavati B, Ozcan M, Piccaglia R. Composition and antimicrobial properties of *Satureja cuneifolia* Ten. and *Thymbra sintenisii* Bornm. et Aznav. subsp. *isaurica* P.H. Davis essential oils. Ann Microbiol. 2004;54(4):393–401.
77. Azaz D, Demirci F, Satil F, Kurkcuoglu M, Baser K. Antimicrobial activity of some *Satureja* essential oils. Z Naturforsch. 2002;57c:817–21. http://dx.doi.org/10.1515/znc-2002-9-1011.
78. Sefidkon F, Jamzad Z. Essential oil omposition of *Satureja spicigera* (C. Koch) Boiss. from Iran. Flav Fragr J. 2004;19:571–3. http://dx.doi.org/10.1002/ffj.1357.
79. Oke F, Aslim B, Ozturk S, Altundag S. Essential oil composition, antimicrobial and antioxidant activities of *Satureja cuneifolia* Ten. Food Chem. 2009;112:874. http://dx.doi.org/10.1016/j.foodchem.2008.06.061.
80. Akgül A, Ozcan M, Chialva F, Monguzzi F. Essential oils of four Turkish wild-growing Labiatae herbs: *Salvia cryptantha* Montbr. et Auch., *Satureja cuneifolia* Ten., *Thymbra spicata* L and *Thymus cilicicus* Boiss. et Bal. J Essent Oil Res. 1999;11:209–14. http://dx.doi.org/10.1080/10412905.1999.9701113.
81. Konakchiev A, Tsankova E. The essential oils of *Satureja montana* ssp. kitaibelii Wierzb. and *Satureja pilosa* var. pilosa Velen from Bulgaria. J Essent Oil Res. 2002;14:120–1. http://dx.doi.org/10.1080/10412905.2002.9699791.
82. Adiguzel A, Ozer H, Kilic H, Cetin B. Screening of antimicrobial of essential oil and methanol extract of *Satureja hortensis* on foodborne bacteria and fungi. Czech J Food Sci. 2007;25:81–9.
83. Gohari AR, Hadjiakhoondi A, Sadat-Ebrahimi E, Saeidnia S, Shafiee A Composition of the volatile oils of *Satureja spicigera* C. Koch Boiss. and S. macrantha C.A. Mey from Iran. Flav Fragr J. 2006;21:510–2. http://dx.doi.org/10.1002/ffj.1613.
84. Gohari AR, Hadjiakhoondi A, Shafiee A, Ebrahimi ES, Mozaffarian V. Chemical composition of the essential oils of *Satureja atropatana* and *Satureja mutica* growing wild in Iran. J Essent Oil Res. 2005;17:17–8.
85. Skocibusic M, Bezic N, Dunkic V. Chemical composition and antidiarrhoeal activities of winter savory (*Satureja montana* L.) essential oil. Pharm Biol. 2003;41(8):622–6. http://dx.doi.org/10.1080/13880200390502180.
86. Gulluce M, Sokmen M, Daferera D, Agar G, Ozkan H, Kartal N, et al. *In vitro* antibacterial, antifungal, and antioxidant activities of the essential oil and methanol extracts of herbal parts and callus cultures of *Satureja hortensis* L. J Agric Food Chem. 2003;51:3958–65. http://dx.doi.org/10.1021/jf0340308.
87. Radonic A, Milos M. Chemical composition and *in vitro* evaluation of antioxidant effect of free volatile compounds from *Satureja montana* L. Free Radical Res. 2003;37(6):673–9. http://dx.doi.org/10.1080/1071576031000105643.

88. Gohari AR, Nourbakhsh MS, Saeidnia S. Comparative investigation of the volatile oils of *Satureja sahendica* Bornm., extracted via steam and hydro distillations. J Essent Oil Bear Pl. 2011;14(6):751–4. http://dx.doi.org/10.1080/0972060X.2011.10643999.
89. Azaz AD, Kürkcüoglu M, Satil F, Baser KHC, Tümen G. *In vitro* antimicrobial activity and chemical composition of some *Satureja* essential oils. Flav Fragr J. 2005;20:587–59. http://dx.doi.org/10.1002/ffj.1492.
90. Sefidkon F, Ahmadi S. Essential oil of *Satureja khuzistanica* Jamzad. J Essent Oil Res. 2000;12:427–8. http://dx.doi.org/10.1080/10412905.2000.9699556.
91. Sefidkon F, Akbari-nia A. Essential oil content and composition of *Satureja sahendica* Bornm. At different stages of plant growth. J Essent Oil Res. 2009;21:112–4. http://dx.doi.org/10.1080/10412905.2009.9700126.
92. Tümen G, Baser KHC. Essential oil of *Satureja spicigera* (C. Koch) Boiss. from Turkey. J Essent Oil Res. 1996;8:57–8. http://dx.doi.org/10.1080/10412905.1996.9700554.
93. Sefidkon F, Jamzad Z. Essential oil analysis of Iranian *Satureja edmondi* and *S. isophylla*. Flav Fragr J. 2006;21:230–3. http://dx.doi.org/10.1002/ffj.1562.
94. Lopez-Reyes JG, Spadaro D, Gullino ML, Garibaldi A. Efficacy of plant essential oils on postharvest control of rot caused by fungi on four cultivars of apples *in vivo*. Flav Fragr J. 2010;25:171–7. http://dx.doi.org/10.1002/ffj.1989.
95. Sefidkon F, Jamzad Z, Mirza M. Chemical variation in the essential oil of *Satureja sahendica* from Iran. Food Chem. 2004;88:325–8. http://dx.doi.org/10.1016/j.foodchem.2003.12.044.
96. Slavkovska V, Jancic R, Bojovic S, Milosavljevic S, Djokovic D. Variability of essential oils of *Satureja montana* L. and *Satureja kitaibelii* Wierzb. ex Heuff. from the central part of the Balkan peninsula. Phytochemistry. 2001;57:71–6. http://dx.doi.org/10.1016/S0031-9422(00)00458-1.
97. Vagionas K, Graikou K, Ngassapa O, Runyoro D, Chinou I. Composition and antimicrobial activity of the essential oils of three *Satureja* species growing inTanzania. Food Chem. 2007;103:319–24. http://dx.doi.org/10.1016/j.foodchem.2006.07.051.
98. Dambolena SJ, Zunino PM, Lucini IE, Zygadlo AJ, Rotman A, Ahumada O, Biurrun F. Essential oils of plants used in home medicine in north of Argentina. J Essent Oil Res. 2009;21:405–9. http://dx.doi.org/10.1080/10412905.2009.9700204.
99. Viturro CI, Molina A, Guy I, Charles B, Guinaudeau H, Fournet A Essential oils of *Satureja boliviana* and *S. parvifolia* growing in the region of Jujuy, Argentina. Flav Fragr J. 2000;15:377–82. http://dx.doi.org/10.1002/1099-1026(200011/12)15:6<377::AID-FFJ926>3.3.CO;2-G.
100. Zygadlo JA, Grosso NR. Comparative study of the antifungal activity of essential oils from aromatic plants growing wild in the central region of Argentina. Flav Fragr J. 1995;10:113–8. http://dx.doi.org/10.1002/ffj.2730100210.
101. Lawrence BM. Labiatae oils-mother nature's chemical factory. In: Lawrence BM, editor. Essential oils carol stream. IL: Allured Publ. Corp.; 1992.
102. Lawrence MB, Bromstein AC. Terpenoides in *Satureja douglasii*. Phytochemistry. 1974;13:1041. http://dx.doi.org/10.1016/S0031-9422(00)91441-9.
103. Karabay–Yavasoglu NU, Baykan S, Ozturk B, Apaydin S, Tuglular I. Evaluation of the antinociceptive and anti-inflammatory activities of *Satureja thymbra* L. essential oil. Pharm Biol. 2006;44(8):585–91.
104. Palic R, Simic N, Andelkovic S, Stojanovic G. Composition of essential oil of selected Balkan's Satureja species and chemotaxonomic implications. J Essent oil Bear Pl. 1998;1:66–81.
105. Tabanca N, Kürkcüoglu M, Baser KHC, Tümen G, Duman H. Composition of the essential oils of *Satureja spinosa* L. J Essent Oil Res. 2004;16:127–8. http://dx.doi.org/10.1080/10412905.2004.9698671.
106. Tarikua Y, Hymeteb A, Hailuc A, Rohloff J. Essential-oil composition, antileishmanial, and toxicity study of *Artemisia abyssinica* and *Satureja punctata* ssp. punctata from Ethiopia. Chem Biodivers. 2010;7:1009–18. http://dx.doi.org/10.1002/cbdv.200900375.
107. Ortet R, Regalado EL, Thomas OP, Pino JA, Fernandez MD. Composition and antioxidant properties of the essential oil from the endemic Cape Verdean *Satureja forbesii*. GIFC. 2010:CP-57.

108. Rojas L, Usubillaga A. Composition of the essential oil of *Satureja brownei* (SW.) Briq. from Venezuela. Flav Fragr J. 2000;15:21–2. http://dx.doi.org/10.1002/(SICI)1099-1026(200001/02)15:1 <21::AID-FFJ861 >3.0.CO;2-G.
109. Palic R, Simic N, Andelkovic S, Stojanovic G. Composition of essential oil of selected Balkan's *Satureja* species and chemotaxonomic implications. J Essent Oil Bear Pl. 1998;1:66–81.
110. Tzakou O, Skaltsa H. Composition and antibacterial activity of the essential oil of *Satureja parnassica* subsp. parnassica. Planta Med. 2003;69:282–4. http://dx.doi.org/10.1055/s-2003-38487.
111. Malagon O, Vila R, Iglesias J, Zaragoza T, Canigueral S. Composition of the essential oils of four medicinal plants from *Ecuador*. Flav Fragr J. 2003;18:527–31. http://dx.doi.org/10.1002/ffj.1262.
112. Habibi Z, Sedaghat S, Ghodrati T, Masoudi S, Rustayian A. Volatile constituents of *Satureja isophylla* and *S. cuneifolia* from Iran. Chem Nat Compd. 2007;43(2):719–21. http://dx.doi.org/10.1007/s10600-007-0244-5.
113. Ezer N, Vila R, Cañigueral S, Adzet T. Essential oil of *Satureja wiedemanniana* (Lallem.) Velen. J Essent Oil Res. 1995;7:91–3. http://dx.doi.org/10.1080/10412905.1995.9698473.
114. Tumen G, Baser KHC, Demirci B, Ermin N. The essential oils of *Satureja coerulea* Janka and *Thymus aznavourii* Velen. Flav Fragr J. 1998;13(1):65–7. http://dx.doi.org/10.1002/(SICI)1099-1026(199801/02)13:1<65::AID-FFJ695>3.0.CO;2-Q.
115. Kotan R, Cakir A, Dadasoglu F, Aydin T, Cakmakci R, Ozer H, et al. Antibacterial activities of essential oils and extracts of Turkish *Achillea*, *Satureja* and *Thymus* species against plant pathogenic bacteria. J Sci Food Agric. 2010;90:145–60. http://dx.doi.org/10.1002/jsfa.3799.
116. Sarac N, Ugur A. Antimicrobial activities of the essential oils of *Origanum onites* L., *Origanum vulgare* L. subspecies hirtum (Link) Ietswaart, *Satureja thymbra* L., and *Thymus cilicicus* Boiss. & Bal. growing wild in Turkey. J Med Food. 2008;11(3):568–73. http://dx.doi.org/10.1089/jmf.2007.0520.
117. Sahin F, Karaman I, Güllüce M, Ogütcü H, Sengül M, Adigüzel A, Oztürk S, et al. Evaluation of antimicrobial activities of *Satureja hortensis* L. J Ethnopharmacol. 2003;87(1):61–5. http://dx.doi.org/10.1016/S0378-8741(03)00110-7.
118. Carraminana JJ, Rota C, Burillo J, Herrera A. Antibacterial efficiency of Spanish *Satureja montana* essential oil against *Listeria monocytogenes* among natural flora in minced pork. J Food Prot. 2008;71(3):502–8.
119. Oliveira T, Soares R, Ramos E, Cardoso M, Alves E, Piccoli R. Antimicrobial activity of *Satureja montana* L. essential oil against *Clostridium perfringens* type A inoculated in mortadella-type sausages formulated with different levels of sodium nitrite. Int J Food Microbiol. 2011;144:546–55. http://dx.doi.org/10.1016/j.ijfoodmicro.2010.11.022.
120. Ghasemi Pirbalouti A, Moalem E. Variation in antibacterial activity of different ecotypes of *Satureja khuzestanica* Jamzad, as an Iranian endemic plant. Indian J Tradit Know. 2013;12(4):623–9.
121. Oussalah M, Caillet S, Lacroix M. Mechanism of action of *Spanish oregano*, *Chinese cinnamon*, and savory essential oils against cell membranes and walls of Escherichia coli O157:H7 and Listeria monocytogenes. J Food Prot. 2006;69(5):1046–55.
122. Cox SD, Gustafson JE, Mann CM, Markham JL, Liew YC, Hartand RP, et al. Tea tree oil causes K1 leakage and inhibits respiration in *Escherichia coli*. Lett Appl Microbiol. 1998;26:355–8. http://dx.doi.org/10.1046/j.1472-765X.1998.00348.x.
123. Helander IM, Alakomi HL, Latva-Kala K, Mattila-Sandholm T, Pol I, Smid EJ, et al. Characterization of the action of selected essential oil components on gram-negative bacteria. J Agric Food Chem. 1998;46:3590–5. http://dx.doi.org/10.1021/jf980154m.
124. Ultee A, Kets EPW, Smid EJ. Mechanisms of action of carvacrol on the food-borne pathogen *Bacillus cereus*. Appl Environ Microbiol. 1999;65:4606–10.
125. Debenedetti S, Muschietti L, van Baren C, Clavin M, Broussalis A, Martino V, Houghton PJ, Warhurst D, Steele J. *In vitro* antiplasmodial activity of extracts of Argentinian plants. J Ethnopharmacol. 2002;80:163–6. http://dx.doi.org/10.1016/S0378-8741(02)00024-7.

126. Egly Feresin G, Tapia A, Lopez NS, Zacchino AS. Antimicrobial activity of plants used in traditional medicine of San Juan province, Argentine. J Ethnopharmacol. 2001;78:103–7. http://dx.doi.org/10.1016/S0378-8741(01)00322-1.
127. Chao S, Young G, Oberg C, Nakaoka K. Inhibition of methicillin-resistant *Staphylococcus aureus* (MRSA) by essential oils. Flav Fragr J. 2008;23:444–9. http://dx.doi.org/10.1002/ffj.1904.
128. Nedorostova L, Kloucek P, Kokoska L, Stolcova M, Pulkrabek J. Antimicrobial properties of selected essential oils in vapour phase against foodborne bacteria. Food Control. 2009;20:157–60. http://dx.doi.org/10.1016/j.foodcont.2008.03.007.
129. Oussalah M, Caillet S, Saucier L, Lacroix M. Inhibitory eVects of selected plant essential oils on the growth of four pathogenic bacteria: E. coli O157:H7, *Salmonella Typhimurium*, *Staphylococcus aureus* and *Listeria monocytogenes*. Food Control. 2007;18:414–20. http://dx.doi.org/10.1016/j.foodcont.2005.11.009.
130. Rota C, Carraminana JJ, Burrilo J, Herrera A. *In vitro* antimicrobial activity of essential oils from aromatic plants against selected food born pathogens. J Food Prot. 2004;67(6):1252–6.
131. Panizzi L, Flamini G, Cioni PL, Morelli I. Composition and antimicrobial properties of essential oils of four Mediterranean Lamiaceae. J Ethnopharmacol. 1993;39:167–70. http://dx.doi.org/10.1016/0378-8741(93)90032-Z.
132. Piccaglia R, Marottil M, Giovanellil E, Deans SG, Eaglesham E. Antibacterial and antioxidant properties of mediterranean aromatic plants. Ind Crop Prod. 1997;2:47–50. http://dx.doi.org/10.1016/0926-6690(93)90010-7.
133. Skocibusic M, Bezic N, Dunkic V. Phytochemical analysis and *in vitro* antimicrobial activity of two *Satureja* species essential oils. Phytother Res. 2004;18(12):967–70. http://dx.doi.org/10.1002/ptr.1489.
134. Hernandez EN, Tereschuk ML, Abdala LR. Antimicrobial activity oflavonoids in medicinal plants from Tafi del Valle (Tucuman, Argentina). J Ethnopharmacol. 2000;73:317–22. http://dx.doi.org/10.1016/S0378-8741(00)00295-6.
135. Caceres A, Alvarez A, Ovando A, Samayoa B. Plants used in Guatemala for the treatment of resprratory diseases. 1. Screening of 68 plants against gram-positive bacteria. J Ethnopharmacol. 1991;31:193–208. http://dx.doi.org/10.1016/0378-8741(91)90005-X.
136. Amanlo M, Fazeli MR, Arvin A, Amin HG, Farsam H. Antimicrobial activity of crude methanolic extract of *Satureja khuzistanica*. Fitoterapia. 2004;75:768–70. http://dx.doi.org/10.1016/j.fitote.2004.09.005.
137. Hosainzadegan H, Delfan B. Evaluation of antibiofilm of dentol. Acta Med Iran. 2009;47(1):35–40.
138. Sadeghi Ghotbabadi F, Alizadeh A, Zadehbagheri M, Kamelmanesh MM, Shaabani M. Phytochemical composition of the essential oil, total phenolic content, antioxidant and antimicrobial activity in Iranian *Satureja sahendica* Bornm. At different ontogenesis conditions. J Med Plants Res. 2012;6(19):3525–34. http://dx.doi.org/10.5897/jmpr11.374.
139. Muller-Riebau F, Berger B, Yegen O. Chemical composition and fungitoxic properties to phytopathogenic fungi of essential oils of selected aromatic plants growing wild in Turkey. J Agric Food Chem. 1995;43:2262–6. http://dx.doi.org/10.1021/jf00056a055.
140. Cristani M, D'Arrigo M, Mandalari G, Castelli F, Sarpietro MG, Micieli D, et al. Interaction of four monoterpenes contained in essential oils with model membranes: implications for their antibacterial activity. J Agric Food Chem. 2007;55:6300–8. http://dx.doi.org/10.1021/jf070094x.
141. Boyraz N, Ozcan M. Antifungal effect of some spice hydrosols. Fitoterapia. 2005;76:661–5. http://dx.doi.org/10.1016/j.fitote.2005.08.016.
142. Behravan J, Ramezani M, Kasaian J, Sabeti Z. Antimycotic activity of the essential oil of *Satureja mutica* Fisch & C.A. Mey from Iran. Flav Fragr J. 2004;19:421–3. http://dx.doi.org/10.1002/ffj.1328.
143. Sekine T, Sugano M, Majid A, Fujii Y. Antifungal effects of volatile compounds from black Zira (*Bunium persicum*) and other spices and herbs. J Chem Ecol. 2007;33:2123–32. http://dx.doi.org/10.1007/s10886-007-9374-2.

144. Sagdic O, Ozturk I, Bayram O, Kesmen Z, Tahsin Yilmaz M. Characterization of butter spoiling yeasts and their inhibition by some spices. J Food Sci. 2010;75(9):597–603. http://dx.doi.org/10.1111/j.1750-3841.2010.01871.x.

145. Giordani R, Regli P, Kaloustian J, Mikail C, Abou L, Portugal H. Antifungal effect of variuos essential oils against *Candida albocans*. potentiation of antifungal action of amphotericine B by essential oil from *Thymus vulgaris*. Phytother Res. 2004;18:990–5. http://dx.doi.org/10.1002/ptr.1594.

146. Dikbas N, Kotan R, Dadasoglu F, Sahin F. Control of *Aspergillus flavus* with essential oil and methanol extract of *Satureja hortensis*. Int J Food Microbiol. 2008;124:179–82. http://dx.doi.org/10.1016/j.ijfoodmicro.2008.03.034.

147. Razzaghi-Abyaneh M, Shams-Ghahfarokhi M, Yoshinari T, Rezaee MB, Jaimand K, Nagasawa H, Sakuda S. Inhibitory effects of *Satureja hortensis* L. essential oil on growth and aflatoxin production by *Aspergillus parasiticus*. Int J Food Microbiol. 2008;123:228–33. http://dx.doi.org/10.1016/j.ijfoodmicro.2008.02.003.

148. Fraternale D, Giamperi L, Bucchini A, Ricci D, Epifano F, Genovese S, Curini M. Chemical composition and antifungal activity of the essential oil of *Satureja montana* from central Italy. Chem Nat Compd. 2007;43(5):622–24. http://dx.doi.org/10.1007/s10600-007-0210-2.

149. Sadeghi-Nejad B, Shiravi F, Alinejadi M, Zarrin M. Antifungal activity of *Satureja khuzestanica* (Jamzad) leaves extracts Jundishapur. J Microbiol. 2010;3(1):36–40.

150. Yamasaki K, Nakano M, Kawahata T, Mori H, Otake T, Ueba N, et al. Anti-HIV-1 activity of herbs in Labiatae. Biol Pharm Bull. 1998;21(8):829–33. http://dx.doi.org/10.1248/bpb.21.829.

151. Abad MJ, Bermejo P, Gonzales E, Iglesias I, Irurzun A, Carrasco L. Antiviral activity of Bolivian plant extracts. Gen Pharmacol. 1999;32(4):499–503. http://dx.doi.org/10.1016/S0306-3623(98)00214-6.

152. Gohari AR, Saeidnia S, Kiuchi F, Honda G. Trypanocidal activity of some endemic species of *Satureja* in Iran. Iran J Pharm Res. 2004;3(2):72–3.

153. Aslan I, Ozbek H, Calmasur O, Sahin F. Toxicity of essential oil vapours to two greenhouse pests, *Tetranychus urticae* Koch and *Bemisia tabaci* Genn. Ind Crop Prod. 2004;19:167–73. http://dx.doi.org/10.1016/j.indcrop.2003.09.003.

154. Isman MB, Wan AJ, Passreiter CM. Insecticidal activity of essential oils to the tobacco cutworm, *Spodoptera litura*. Fitoterapia. 2001;72:65–8. http://dx.doi.org/10.1016/S0367-326X(00)00253-7.

155. Pavela R. Larvicidal property of essential oils against culex quinquefasciatus Say (Diptera: culicidae). Ind Crop Prod. 2009;30:311–5. http://dx.doi.org/10.1016/j.indcrop.2009.06.005.

156. Karpouhtsis I, Pardali E, Feggou E, Kokkini S, Scouras ZG, Mavragani-Tsipidou P. Insecticidal and genotoxic activities of oregano essential oils. J Agric Food Chem. 1998;46:1111–5. http://dx.doi.org/10.1021/jf970822o.

157. Lindberg Madsen H, Serensen B, Skibsted LH, Bertelsen G. The antioxidative activity of summer savory (*Satureja hortensis* L.) and rosemary (*Rosmarinus oflcinalis* L.) in dressing stored exposed to light or in darkness. Food Chem. 1998;63(2):173–80. http://dx.doi.org/10.1016/S0308-8146(98)00038-7.

158. Marinova EM, Yanishlieva NV. Antioxidative activity of extracts from selected species of the family Lamiaceae in sunflower oil. Food Chem. 1997;58(3):245–8.

159. Dorman HJD, Hiltunen R. Fe(III) reductive and free radical-scavenging properties of summer savory (*Satureja hortensis* L.) extract and subfractions. Food Chem. 2004;88:193–9. http://dx.doi.org/10.1016/j.foodchem.2003.12.039.

160. Deighton N, Glidewell MS, Deans SG, Goodman AB. Identification by EPR spectroscopy of carvacrol and thymol as the major sources of free radicals in the oxidation of plant essential oils. J Sci Food Agric. 1993;63:221–5. http://dx.doi.org/10.1002/jsfa.2740630208.

161. Yanishlieva N, Marinova MM, Marekov NI, Gordon HM. Effect of an ethanol extract from summer savory (*Saturejae hortensis* L) on the stability of sunflower oil at frying Temperature. J Sci Food Agric. 1997;74:524–30. http://dx.doi.org/10.1002/(SICI)1097-0010(199708)74:4<524::AID-JSFA829>3.0.CO;2-W.

162. Giao MS, Leitao I, Pereira A, Borges AB, Guedes CJ, Fernandes JC, Belo L, Santos-Silva A, Hogg TA, Pintado ME, Malcat F. Plant aqueous extracts: antioxidant capacity via haemolysis and bacteriophage P22 protection. Food Control. 2010;21:633–8. http://dx.doi.org/10.1016/j.foodcont.2009.08.014.
163. Grosso C, Oliveira AC, Mainar AM, Urieta JS, Barroso JG, Palavra AMF. Antioxidant activities of the supercritical and conventional *Satureja montana* extracts. J Food Sci. 2009;74(9):713–7. http://dx.doi.org/10.1111/j.1750-3841.2009.01376.x.
164. Cavar S, Maksimivic M, Solic ME, Jerkovic-Mujkic A, Besta R. Chemical composition and antioxidant and antimicrobial activity of two *Satureja* essential oils. Food Chem. 2008;111:648–53. http://dx.doi.org/10.1016/j.foodchem.2008.04.033.
165. Prieto MJ, Iacopini P, Cioni P, Chericoni S. *In vitro* activity of the essential oils of *Origanum vulgare, Satureja montana* and their main constituents in peroxynitrite-induced oxidative processes. Food Chem. 2007;104:889–95. http://dx.doi.org/10.1016/j.foodchem.2006.10.064.
166. Mohammadi Motamed S, Naghibi F. Antioxidant activity of some edible plants of the Turkmen Sahra region in northern Iran. Food Chem. 2010;119:1637–42. http://dx.doi.org/10.1016/j.foodchem.2009.09.057.
167. Ozkan G, Simsek B, Kuleasan H. Antioxidant activities of *Satureja cilicica* essential oil in butter and *in vitro*. J Food Eng. 2007;79:1391–6. http://dx.doi.org/10.1016/j.jfoodeng.2006.04.020.
168. Caillet S, Yu H, Lessard S, Lamoureux G, Ajdukovic D, Lacroix M. Fenton reaction applied for screening natural antioxidants. Food Chem. 2007;100:542–52. http://dx.doi.org/10.1016/j.foodchem.2005.10.009.
169. Angelini GL, Carpanese G, Cioni LP, Morelli I, Macchia M, Flamini G. Essential oils from mediterranean Lamiaceae as weed germination inhibitors. J Agri Food Chem. 2003;51:6158–64. http://dx.doi.org/10.1021/jf0210728.
170. Mazzio EA, Soliman K. *In vitro* screening for the tumoricidal properties of international medicinal herbs. Phytother Res. 2009;23:385–98. http://dx.doi.org/10.1002/ptr.2636.
171. Lin LT, Liu LT, Chiang LC, Lin CC. *In vitro* anti-hepatoma activity of fifteen natural medicines from Canada. Phytother Res. 2002;16:440–4. http://dx.doi.org/10.1002/ptr.937.
172. Giao M, Gonzalez-Sanjose ML, Muniz P, Rivero-Prez MD, Kosinska M, Pintado M, Malkata F. Protection of deoxyribose and DNA from degradation by using aqueous extracts of several wild plants. J Sci Food Agric. 2008;88:633–40. http://dx.doi.org/10.1002/jsfa.3128.
173. Christensen KB, Minet A, Svenstrup H, Grevsen K, Zhang H, Schrader E, Rimbach G, Wein S, Wolffram S, Christiansen K, Christensen LP. Identification of plant extracts with potential antidiabetic properties: effect on human peroxisome proliferator-activated receptor (PPAR) and isulin-stimulated glucose uptake. Phytother Res. 2009;23:1316–25.
174. Schrader E, Wein S, Kristiansen K, Christensen LP, Rimbach G, Wolffram S. Plant extracts of winter savory, purple coneflower, buckwheat and black elder activate PPAR-γ in vitro but do not exert anti-diabetic activity in db/db mice *in vivo*. Plant Foods Hum Nutr. 2012;67(4):377–83. http://dx.doi.org/10.1007/s11130-012-0322-0.
175. Saadat M, Pournourmohammadi S, Donyavi M, Khorasani R, Amin G, Salehnia AN, Abdollahi M. Alteration of rat hepatic glycogen phosphorylase and phosphoenolpyruvate carboxykinase activities by *Satureja khuzestanica* Jamzad essential oil. J Pharm Pharm Sci. 2004;7(3):327–31.
176. Abdollahi M, Salehnia A, Mortazavi SH, Ebrahimi M, Shafiee A, Fouladian F, et al. Antioxidant, antidiabetic, antihyperlipidemic, reproduction stimulatory properties and safety of essential oil of *Satureja khuzestanica* in rat *in vivo*: a oxicopharmacological study. Med Sci Monit. 2003;9:331–5.
177. Slanc P, Doljak B, Kreft S, Lunder M, Janes D, Strukelj B. Screening of selected food and medicinal plant extrcats for pancreatic lipase inhibition. Phytother Res. 2009;23:874–7. http://dx.doi.org/10.1002/ptr.2718.

178. Mchedlishvili D, Kuchukashvili Z, Tabatadze TGD. Influence of flavonoids isolated from *Satureja hortensis* L. on hypercholesterolemic rabbits. Indian J Pharmacol. 2005;37:259–60. http://dx.doi.org/10.4103/0253-7613.16577.
179. Vosough-Ghanbari S, Rahimi R, Kharabaf S, Zeinali S, Mohammadirad A, Amini S, Yasa N, Salehnia A, Toliat T, Nikfar S, Larijani B, Abdollahi M. Effects of *Satureja khuzestanica* on serum glucose, lipids and markers of oxidative stress in patients with type 2 diabetes mellitus: a double-blind randomized controlled trial. eCAM. 2010;7(4):465–70.
180. Loizzo MR, Saab MA, Tundis R, Menichini F, Piccolo V, Statti AG, et al. *In vitro* inhibitory activities of plants used in Lebanon traditional medicine against angiotensin converting enzyme (ACE) and digestive enzymes related to diabetes. J Ethnopharmacol. 2008;119:109–16. http://dx.doi.org/10.1016/j.jep.2008.06.003.
181. Silva F, Martins A, Salta J, Neng RN, Nogueira J, Mira D, Gaspar N, Justino J, Grosso C, Urieta J, Palavra A, Rauter A. Phytochemical profile and anticholinesterase and antimicrobial activities of supercritical versus conventional extracts of *Satureja montana*. J Agric Food Chem. 2009;57(24):11557–63. http://dx.doi.org/10.1021/jf901786p.
182. Rojas V, Ortega T, Villar A. Pharmacological activity of the extracts of Two *Satureja obovata* varieties on isolated smooth muscle preparations. Phytother Res. 1994;8:212–7. http://dx.doi.org/10.1002/ptr.2650080405.
183. Rojas V, Somoza B, Ortega T, Villar A. Isolation of vasodilatory active flavonoids from the traditional remedy *Satureja obovata*. Planta Med. 1996;62(3):272–4. http://dx.doi.org/10.1055/s-2006-957876.
184. Rojas V, Somoza B, Ortega T, Villar A. Different mechanisms involved in the vasorelaxant effect of flavonoids isolated from *Satureja obovata*. Planta Med. 1996;62(6):554–6. http://dx.doi.org/10.1055/s-2006-957969.
185. Rojas V, Somoza B, Ortega T, Villar A, Tejerina T. Vasodilatory effect in rat aorta of eriodictyol obtained from *Satureja obovata*. Planta Med. 1999;65(3):234–8. http://dx.doi.org/10.1055/s-1999-13986.
186. Morteza-Semnani K, Saeedi M, Hamidian M, Vafamehr H, Dehpour AR. Anti-inflammatory, analgesic activity and acute toxicity of *Glaucium grandiflorum* extract. J Ethnopharmacol. 2002;80:181–6. http://dx.doi.org/10.1016/S0378-8741(02)00027-2.
187. Tjolsen A, Berge OG, Hunskaar S, Rosland JH, Hole K. The formalin test: an evaluation of the method. Pain. 1992;51:5–17. http://dx.doi.org/10.1016/0304-3959(92)90003-T.
188. Hajhashemi V, Ghannadi A, Pezeshkian SK. Antinociceptive and anti-inflammatory effects of *Satureja hortensis* L. extracts and essential oil. J Ethnopharmacol. 2002;82:83–7. http://dx.doi.org/10.1016/S0378-8741(02)00137-X.
189. Gonzales E, Iglesias I, Carretero E, Villar A. Gastric cytoprotection of Bolivian medicinal plants. J Ethnopharmacol. 2000;70:329–33. http://dx.doi.org/10.1016/S0378-8741(99)00183-X.
190. Aydin S, Ozturk Y, Beis R, Can Bager KH. Investigation of *Origanum onites, Sideritis congesta* and *Satureja cuneifolia* essential oils for analgesic activity. Phytother Res. 1996;10:342–4. http://dx.doi.org/10.1002/(SICI)1099-1573(199606)10:4<342::AID-PTR832>3.0.CO;2-W.
191. Ghazanfari G, Minaie B, Yasa N, Ashtaral Nakhai L, Mohammadirad A, Nikfar S, Dehghan G, Shetab Boushehri V, Jamshidi H, Khorasani R, Salehnia A, Abdollahi M. Biochemical and histopathological evidences for beneficial effects of *Satureja khuzestanica* Jamzad essential oil on the mouse model of inflammatory bowel diseases. Toxico Mech Methods. 2006;16:365–72. http://dx.doi.org/10.1080/15376520600620125.
192. Uslu C, Karasen RM, Sahin F, Taysi S, Akcay F. Effects of aqueous extracts of *Satureja hortensis* L. on rhinosinusitis treatment in rabbit. J Ethnopharmacol. 2003;88:225–8. http://dx.doi.org/10.1016/S0378-8741(03)00236-8.
193. Haeri S, Minaie B, Amin G, Nikfar S, Khorasani R, Esmaily H, Salehnia A, Abdollahi M. Effect of *Satureja khuzestanica* essential oil on male rat fertility. Fitoterapia. 2006;77:495–9. http://dx.doi.org/10.1016/j.fitote.2006.05.025.
194. Rezvanfar MA, Sadrkhanlou RA, Ahmadi A, Shojaei-Sadee H, Rezvanfar MA, Mohammadirad A, Salehnia A, Abdollahi M. Protection of cyclophosphamide-induced toxicity in reproductive tract histology, sperm characteristics, and DNA damage by an herbal source;

evidence for role of free-radical toxic stress. Hum Exp Toxicol. 2008;27:901–10. http://dx.doi.org/10.1177/0960327108102046.

195. Rezvanfar MA, Farshid AA, Sadrkhanlou RA, Rezvanfar MA, Salehnia A, Abdollahi M. Benefit of *Satureja khuzestanica* in subchronically rat model of cyclophosphamide-induced hemorrhagic cystitis. Exp Toxicol Pathol. 2010;62:323–30. http://dx.doi.org/10.1016/j.etp.2009.05.005.
196. Basiri S, Esmaily H, Vosough-Ghanbari S, Mohammadirad A, Yasa N, Abdollahi M. Improvement by *Satureja khuzestanica* essential oil of malathion-induced red blood cells acetylcholinesterase inhibition and altered hepatic mitochondrial glycogen phosphorylase and phosphoenolpyruvate carboxykinase activities. Pestic Biochem Physiol. 2007;89:124–9. http://dx.doi.org/10.1016/j.pestbp.2007.04.006.
197. Mosaffa F, Behravan J, Karim G, Iranshahi M. Antigenotoxic effects of *Satureja hortensis* L. on rat lympho-cytes exposed to oxidative stress. Arch Pharm Res. 2006;29(2):159–64.
198. Rechinger KH. Flora iranica. Labiatae. 150 Vol. Austria: Academische Druck-U-Verganstalt; 1986.pp. 495–4.
199. Mozaffarian V. A dictionary of Iranian plant names. Tehran: Farhang Moaser Publication; 1996. pp. 483–4.
200. Gohari AR, Saeidnia S, Hadjiakhoondi A. Trypanocidal activity of the essential oil of *Satureja macrantha* and its volatile components. Int J Essent Oil Ther. 2007;1:184–6.
201. Gohari AR, Saeidnia S, Hadjiakhoondi A, Naghinejad A, Yagura T. Trypanocidal activity of some medicinal plants against the epimastigotes of *Trypanosome cruzi*. J Med Plants. 2008;7:22–6.
202. Zarrin M, Amirrajab N, Sadeghi-Nejad B. *In vitro* antifungal activity of *Satureja Khuzestanica* Jamzad against *Cryptococcus neoformans*. Pak J Med Sci. 2010;26:880–2.
203. Gohari AR, Hadjiakhoondi A, Sadat-Ebrahimi SE, Saeidnia S, Shafiee A. Cytotoxic triterpenoids from *Satureja macrantha* C. A. Mey. Daru. 2005;13: 177–81.
204. Sadeghi-Nejad B, Saki J, Khademvatan S, Nanaei S. *In vitro* antileishmanial activity of the medicinal plant—*Satureja khuzestanica* Jamzad. J Med Plants Res. 2011;5:5912–5.
205. Gohari AR, Hadjiakhoondi A, Sadat-Ebrahimi SE, Saeidnia S, Shafiee A. Composition of volatile oils of *Satureja spicigera* and *Satureja macrantha* from Iran. Flav Frag J. 2005;21:348–50. http://dx.doi.org/10.1002/ffj.1642.
206. Saeidnia S, Nourbakhsh MS, Gohari AR, Davood A. Isolation and identification of the main compounds of *Satureja sahendica* Bornm. Aust J Basic and Appl Sci. 2011;5:1450–3.
207. Gohari AR, Ostad SN, Moradi-Afrapoli F, Malmir M, Tavajohi S, Akbari H, Saeidnia S. Evaluation of the cytotoxicity of *Satureja spicigera* and Its main compounds. ScientificWorld J. 2012;2012:203861. http://dx.doi.org/10.1100/2012/203861.
208. Ahanjan M, Ghaffari J, Mohammadpour G, Nasrolahie M, Haghshenas MR, Mirabi AM. Antibacterial activity of *Satureja bakhtiarica* bung essential oil against some human pathogenic bacteria. Afr J Microb Res. 2011;5:4764–8.
209. Gohari AR, Saeidnia S, Hadjiakhoondi A, Honda G. Isolation and identification of four sterols from Oud. J Med Plants. 2008;7:47–55.
210. Saeidnia S, Moradi-Afrapoli F, Gohari AR, Malmir M. Cytotoxic flavonoid from *Achillea talagonica* Bioss. J Med Plants. 2009;8:52–6.
211. Schwarz K, Ernst H, Ternes W. Evaluation of anti- oxidative constituents from thyme. J Sci Food Agric. 1996;70:217–23. http://dx.doi.org/10.1002/(SICI)1097-0010(199602)70:2<217::AID-JSFA488>3.0.CO;2-Y.
212. Malmir M, Gohari AR, Saeidnia S. Flavonoids from the aerial parts of *Satureja khuzestanica*. Planta Med. 2012;78:PI365. http://dx.doi.org/10.1055/s-0032-1321052.
213. Saeidnia S, Yassa N, Rezaeipoor R, Shafiee A, Gohari AR, Kamalinejad M, Goodarzy S. Immunosuppressive principles from *Achillea talagonica*, an endemic species of Iran. Daru. 2009;17:37–41.
214. Gohari AR, Saeidnia S, Shahverdi AR, Yassa N, Malmir M, Mollazade K, et al. Phytochemistry and antimicrobial compounds of *Hymenocrater calycinus*. Eur Asia J Bio Sci. 2009;3:64–8. http://dx.doi.org/10.5053/ejobios.2009.3.0.9.

215. Lu Y, Foo LY. Rosmarinic acid derivatives from *Salvia officinalis*. Phytochemistry. 1999;51:91–4. http://dx.doi.org/10.1016/S0031-9422(98)00730-4.
216. Sanchez de Rojas VR, Somoza B, Ortega T, Villar AM. Different mechanisms involved in the vasorelaxant effect of flavonoids isolated from *Satureja obovata*. Planta Med. 1996;62:554–6. http://dx.doi.org/10.1055/s-2006-957969.
217. Ruh MF, Zacharewski T, Connor K, Howell J, Chen I, Safe S. Naringenin: a weakly estrogenic bioflavonoid that exhibits antiestrogenic activity. Biochem Pharmacol. 1995;50:1485–93. http://dx.doi.org/10.1016/0006-2952(95)02061-6.
218. van Baren C, Anao I, Leo Lira P D, Debenedetti S, Houghton P, Croft S, Martino V. Triterpenic acids and flavonoids from *Satureja parvifolia*. Evaluation of their antiprotozoal activity. Z Naturforsch. 2006;61c:189–92. http://dx.doi.org/10.1515/znc-2006-3-406.
219. Balali P, Saeidnia S, Soodi M. Protective effects of some medicinal plants from Lamiaceae family against beta-amyloid induced toxicity in PC12 cell. Tehran Univ Med J. 2012;70:402–9.

Printed in the United States
By Bookmasters